EVERYTHING YOU SHOULD KNOW ABOUT THE WORLD'S ENVIRONMENT, BUT ARE INDIFFERENT TO ASK!

By

Brendan James Magee, Esquire

And

A.C. Underwood II

March, 2009

Copyrighted by Brendan James Magee, And A. C. Underwood, II – 2009

All rights reserved

Without limiting the rights under copyright reserved above, no part of this publication may be reproduced, stored in or introduced into retrieval systems, or transmitted, in any form, or by any means (electronic, mechanical, photocopying, recording, or otherwise), without the prior written permission of the copyright owner.

"The definition of insanity is the belief, that you can get different results by doing the same thing."

Einstein

Note: See back of book for Table of Contents.

PREFACE

Our book, Everything You Should Know About The World's Environment, But Are Indifferent To Ask! was written for educational purposes to enlighten the public and provoke serious thought and debate. The work is written by two Americans, who still believe that this country is for the people and by the people. Unfortunately the people are facing a global struggle in the form of environment degradation of epidemic proportions.

Green living and the "greening" of America are recent hot topics, however the major problems have gone largely overlooked. This book addresses such issues in a thoughtful, reasonable, and objective standard and point of view.

A. C. Underwood is a University of Pennsylvania Wharton School of Business graduate, who worked in an executive position for a Fortune 500 company, based in Philadelphia, Pa. for over 36 years, dealing with environmental and pollution issues. Mr. Underwood has authored several books on various subjects, since his retirement from the executive forum.

Mr. Magee is an environmental attorney with special expertise in Environment Law, Science and Policy, including a Certificate of Excellence in Achievement in Environmental Law from his alma mater, University of Pittsburgh, where he earned his Doctor of Jurisprudence. Mr. Magee also carries a Bachelor of Science degree in Environment Science, Engineering and Policy from Drexel University. Mr. Magee is licensed to practice law in the State of Colorado, and has litigated several environment cases in his tenure as an attorney. Mr. Magee has also worked as an environmental scientist for a Fortune 500 company, and has counseled various companies concerning environmental issues.

This book is the culmination of over 40 years of experience in the real world environmental business and law. In the work we seek to provide a litany of environmental issues, the status of environmental law on such issues, and potential solutions going forward. Our purpose is to educate the masses with a firm conviction that an educated electorate is the best possible mode of real change in a democracy.

"You can fool some of the people some of the time
You can even fool all of the people some of the time,
But you cannot fool all of the people all of the time."

      Abraham Lincoln,
16th President of the United States

CHAPTER ONE:
INTRODUCTION

Man's survival on earth is dependent upon three broad categories of elements: land, air and water. The destruction and/or degradation of any one of these three can have disastrous results upon the survival of the human race. Earth is all that man has, our only home. This reality should pervade our collective conscience in a more prominent way than has historically been the case.

Land with too much or too little sun will not produce the crops necessary for survival. Air with too many toxicants can kill, mutate, or otherwise cause serious physiological unseen harm. Water polluted, will not provide for potability, hygiene or recreation.

This book is organized to point out several environmental disasters which have occurred in the past generations and continue to occur throughout the global environment. Moreover, this book will show government's inaction in remedying these issues, and finally provide some thoughtful solutions to certain global environmental issues and problems.

The writing is fact-based and evidence driven. Every proposition is supported by data and evidence. Every source checked and rechecked again and again. The purpose of the book is to educate the populace as to the devastations and outright onslaught against the environment that have occurred and are continuing to take place right under our noses. The contents aren't meant to be political commentary or partisan rhetoric. It is simply the truth.

The book may frighten or outrage the reader, but that reaction is expected, when dealing with the harsh realities of our lives. It is important to understand that planet earth is your home. You wouldn't want someone entering your house and trashing the place from basement to attic. So, why put up with this treatment in every sector you frequent. Pollution is everywhere: your backyard, your hometown, where you visit, your state, your country. You may say that you aren't affected by the global environment, but you are. What is more important, so are your innocent children, who never polluted or caused harm to the environment.

What about them and future generations? Will we leave them without air to breath, water to drink, and land to farm? Is that really the legacy we want to bestow upon them?

In our work we believe and propose to you that we can live in harmony with Earth without drastically changing our lifestyles. The solutions are there, if we want to accept them and commit ourselves to change. We believe that this is no longer an option. The question has now become how much more pollution and abuse can this old planet of ours handle? The world environment has always been in a transitory state, but how much anthropogenic interference is too much. How long before the Earth becomes like the other lifeless planets in our solar system: uninhabitable for mankind? Obviously, man can't survive in our current life form without a life sustaining planet. What is man doing with the one he has: we call it destruction of our own making. Doomsday won't come from God's dictate; mankind will do it on his own!

In Everything You Should Know, etc. you will read about atomic bombs, oil spills, fires, nuclear power plant meltdown, the combustion engine, air pollution, radioactive elements, cancer, the stratospheric ozone hole, water pollutants, land degradation, U. S. environment law, environmental treaties and more. We fit it all in there. See what you can learn from it.

CHAPTER TWO
CONCEPTS OF ECOSYSTEMS

The Earth is comprised of interrelated ecosystems. There are related cycles involving water, carbon, oxygen, nitrogen, solar radiation, phosphorus and sulfur. (1) Disturbances in these systems can injure other parts of the cycle or other systems.

The Earth is powered by the Sun. Solar radiation is absorbed as heat upon penetrating the Earth's atmosphere. The solar heat drives ocean currents and produces water vapors. Winds are created as the heat rises through the atmosphere.

The light spectrum also powers photosynthesis in plants. The sun provides enough power to produce an additional two billion tons of organic matter per year. (2) This is essential to the ecosystem.

Water moves in the hydrologic cycle, is essential for plant and animal life, and is the most abundant substance in the biosphere. Hydrogen is essential to this system, being the most abundant element on Earth. Hydrogen plays key roles in fossil fuels, and will likely have added roles in future alternate energy sources.

More than 90% of the Earth's water is held in the world's oceans, with another 2% making up the polar icecaps. Less than one tenth of one percent of the Earth's water is in freshwater surface systems. (3) Water in liquid state will dissolve into almost any solution, often resulting in serious water pollution. Anthropogenic discharge of waste into water bodies adds suspended pollutants, drastically changing the contents of the fresh water body and negatively affecting the hydrologic cycle.

(1) Environmental Law Examples and Explanations (Steven Ferry, Aspen Publishers, 3rd Ed. 2004)

(2) Environmental Law Examples and Explanations, p. 2.

(3) Environmental Law Examples and Explanations, p.2

Carbon also moves through a cycle in the biosphere, starting as atmospheric carbon dioxide and ending in living organisms such as plants. Carbon makes up the basis of fossil fuels, including oil, coal and natural gas. Carbon dioxide is the number one heat trapping gas that leads to global climate change, directly accelerated through man's use of cars and electricity.

Oxygen also exists in its own cycle, and is primarily extant due to plant photosynthesis. Oxygen exists in the troposphere in a diatomic state, whereas it exists in the stratosphere in a diatomic state known as Ozone. The Ozone layer is essential to life on Earth. However, Ozone can be deadly if inhaled in large quantities, and exists in the troposphere primarily resultant from the burning of fossil fuels. Ozone is a criteria air pollutant in the Clean Air Act, due to its hazardous health nature, which is incongruent with the Act's purpose to protect the quality of the Nation's air, and promote the public health and welfare.

(4) Clean Air Act, sec. 101(b)(1), 42 U. S. C. sec. 1857(b)(1) (1970). See also W. Rogers, Environmental Law sec. 3.1 (1977) and ASARCO v. U. S. Environmental Protection Agency, 578 F.2d 319 (1978)

CHAPTER THREE: DISASTERS

Atomic Bomb:

United States drops bomb on Hiroshima, August 6, 1945

And on Nagasaki, August 9, 1945 (introduction to Radioactive fallout in addition to the bomb's extreme heat and concussion.)

(2) Hiroshima, Japan −155,000 people were killed when the Atomic Bomb was dropped on the Hiroshima community. This figure includes radiation deaths occurring within a year. As a result of the bomb's concussion, over 65% of the city's structures were destroyed or damaged.

(3) Pandemic: An epidemic disease, caused by the germ Bacterium Pasteurella Prestis, ravaged Europe and Asia between 1347–1351. It occurred in various forms i.e. bubonic, pneumonic, and septicemic plague; resulting in 75,000,000 deaths.

(4) Genocide: The Mongol Empire exterminated 35,000,000 Chinese peasantry, between 1311–1340.

(3) Influenza: Is a viral disease mostly

spread through the air by afflicted victims' coughing and sneezing.   An estimated 21,640,000 deaths worldwide resulted from it between 1918 and 1919.

(2) Famine Death Toll: Between 1959–1961 approximately 40 million people died from malnutrition in mainland China. The famine was blamed in part on unsuccessful implementation of new farming methods, which were introduced in Chairman Mao's newly formed People's Republic of China, 1949.

(2) Chemical Disaster: 4,000 people died and thousands more were left with permanent disabilities, when a toxic cloud of Methyl Isocyanate (MIC) gas descended on a community near a Union Carbide pesticide plant in Bhopal, India 12/3/84.

(2) Worst Siege in History: was on Leningrad, USSR (now St. Petersburg, Russia). The city fell under siege by the German army for 880 days (August 30, 1941 to January 27, 1944). Estimated 1.5 million defenders and citizens lost their lives; 641,000 died of starvation and 17,000 by shelling.

(2) Avalanche Disaster: In the Dolomite

Valley, Tyrolean Alps, December 13, 1916 an estimated 40,000 – 80,000 lost their lives during World War I. The avalanches were triggered by gunfire.

(1) Fires: In Durango, CO, June 25, 2002 a fire along Missionary Ridge in the Rocky Mountains (one of a host of wildfires) plagued America's western states in the summer of 2002. The wildfires weren't confined to the west: in a summer struck by drought, big blazes burned in various states. By mid-August, some 56,000 individual wildfires had devastated a total of some 6 million acres of forest.

(1) Fires: April–May, 2000 in northern New Mexico, a prescribed fire started by National Park Service raged out of control, destroying 235 structures and forcing evacuation of more than 20,000 people. The blaze consumed 47,000 acres (estimated), and threatened Los Alamos National Laboratory.

(1) Fires: In the western United States (November 3, 2000) – a combination of hot, dry weather and more than normal dry vegetation led to one of the most destructive forest fire seasons in American history. 7.2 million acres (estimated)

had burned nationwide, nearly double the ten year average. States hardest hit were Alaska, Idaho, Montana, New Mexico and Oregon.

(1) Fires: The Painted Cave fire in Santa Barbara, CA (June, 1990) consumed 4,900 acres and destroyed 641 structures.

(1) Fires: In western United States (August – September, 1988) fires destroyed over 1.2 million acres in Yellowstone Nation Park, and damaged Alaska woodlands.

Source: (1)Time Almanac 2003, Time, Inc. Home Entertainment, With Information Please, Information Please, Part of Family Education Network, Inc. Page 619
(2)Guinness World Records Limited, 2002, Page 112 – 114
(3)Guinness World Records Limited, 2003, Page 59
(4)Funk & Wagnall's Encyclopedia, Standard Works Publishing Co., New York, NY, Page 7107
(5)Funk & Wagnall's Encyclopedia, Page 3938

Note concerning Fires: We are limiting our findings to just five fires; obviously there are more, and every bit as devastating. Some of them are from natural causes, but many, I'm sorry to say, are man-made, i.e. careless campers, cigarettes, on purpose, etc.

Comments about disasters: All of the above disasters prove that mankind has a resiliency for bouncing back in time of disasters and surviving them, coming away stronger and smarter for the experience.   If there is hope presented in this book, it is just that: man is a survivor. However, there is no better time than now for my fellow man to begin dealing with his newest disaster: pollution. As I will show later in the work, there are energy alternatives. Isn't it time to begin using them? The effort doesn't need to be drastic at first, but if steady and determined, would be a good start. Nothing is being done now!

CHAPTER FOUR:

OIL SPILLS – FOSSIL FUEL

Note: I heard a quote about America's oil dependency some years ago, which goes something like this: "Oil is the American's bread." I believe this is true. If it is, then it is a very messy one for the environment, both before and after its' use. Consider the following incidents.

The Argo Merchant ship ran aground, and broke apart southeast of Nantucket Island on December 15, 1976 at Buzzards Bay, MA. The accident resulted in the loss of the ship's entire cargo of 7.7 million gallons of fuel oil.

The super-tanker Amoco Cadiz was wrecked off the coast of Portsall, France on March 16, 1978, spilling 68 million gallons of oil. The mishap, considered the world's largest tanker disaster, caused widespread environmental damage over 100 mile of Brittany coast.

On June 3, 1979, the exploratory oil well Ixtoc 1 blew out, spilling an estimated 140 million gallons of crude oil into the Gulf of Mexico. It is the largest known oil spill, but fortunately it had a low environmental impact.

The Exxon Valdez tanker hit an undersea reef off the Prince William Sound, Alaska on March 24, 1989, releasing 10 million gallons of oil into the water.

An explosion on the Iranian supertanker, the Kharg 5, caused 19 million gallons of crude oil to spill into the Atlantic Ocean. The accident occurred on December 19, 1989 about 400 miles north of Las Palmas the Canary Islands, forming a 100 square mile oil slick.

During the Persian Gulf War on January 25, 1991, Iraq deliberately released an estimated 460 million gallons of crude oil into the Persian Gulf from tankers 10 miles off Kuwait. On January 27, U. S. warplanes successfully bombed the pipe systems to stop the flow of oil.

A Russian dam built to contain oil bursts failed to do so on September 8, 1994. The mishap resulted in the spilling of oil into Kolva River tributary. U. S. Energy Department viewed the loss at about two million barrels. The Russian state-owned oil company disagreed; claiming the spill was only 102,000 barrels. Who do you believe?

(2) An area of the untouched Arctic tundra of the Komi, Republic of Russia was polluted by an oil spill, between August and September, 1994. Known as the Usinsk Accident, the soil covered by the spill totaled 5,213 acres (the size of El Salvador). Thousands of tons flowed from damaged pipes transporting oil from an oil refinery plant. Estimates of the amount of oil lost were placed at 31 million gallons; valued at 10 billion dollars. This is the worst case of land pollution ever recorded.

Off the Welsh coast the supertanker, Sea Empress ran aground at the port of Milford Haven, Wales on February 15, 1996. The damaged tanker spewed out 70,000 tons of crude oil, creating a 25 mile slick.

Source: (1) Time Almanac 2003, Page 623
(2) Guinness World Records, 2002, Page 106

CHAPTER FIVE:
NUCLEAR POWER PLANT ACCIDENTS

Note: The other type of energy, Nuclear, is similar to Oil in that it pollutes two ways: in the form of radioactive fallout and the disposal of its' waste. The following is an example of the former. The latter I will discuss in detail later on in the work.

Chalk River, near Ottawa, Canada (December, 12, 1952) a partial meltdown of the reactor's uranium fuel core occurred after the accidental removal of four control rods. No injuries were reported, even though millions of gallons of radioactive water had accumulated inside the reactor.

On October 7, 1957 a fire in a graphite-cooled reactor at the Windscale Plant near Liverpool, England spewed radiation over the countryside, contaminating a 200 square mile area.

(October 7, 1957) In a Soviet Union nuclear weapons factory, 12 miles from the city of Kyshtym, an explosion of radioactive wastes forced the evacuation of over 10,000 people from the contaminated area. No casualties were reported by the factory's officials.

(1976) A radioactive core of reactor in the Lubmin Nuclear Power Plant on the outskirts of Greifswald, East Germany nearly melted down due to the failure of safety systems during a fire.

(March 28, 1979) In a nuclear power plant located at Three Mile Island near Harrisburg PA, one of two reactors lost its coolant causing overheating and partial meltdown of its uranium core. Some radioactive water and gases were released before the accident could be brought under control.

(April 25, 1986) At the Chernobyl Nuclear Plant, near Kiev, Ukraine an explosion and fire in the graphite core, one of four reactors, released radioactive material that spread over part of the Soviet Union, eastern Europe, Scandinavia, and later western Europe. The accident was the worse to date: 31 were reported dead, total casualties unknown.

(September 30, 1999) At a nuclear plant in Tokaimura, Japan an uncontrolled chain reaction in a uranium-processing factory spewed high levels of radioactive gas into the air killing one worker and seriously injuring two others. It is considered Japan's worst nuclear accident.

Source: Time Almanac 2003, Page 617

CHAPTER SIX

HENRY FORD (1863 –1947)

AND WHAT HAS BEEN DONE TO CLEAN UP HIS CAR?

Henry Ford was an American engineer, who founded the Ford Motor Company at Detroit, MI in 1903. He didn't actually invent the automobile, but pioneered in the standardization and mass production of the vehicle via the assembly line. Using these techniques, he was able to manufacture reliable, low cost cars and other motor vehicles. He produced the 'gasoline buggy' in 1893, and his first farm tractors in 1915.

Source: Webster's Seventh New Collegiate Dictionary, A Merriam-Webster G. & C. Merriam Company, Publishers, Springfield, MA, 1963

Note: Ford's car as far as pollution goes has remained this way until the present with some exceptions: 1. Nonleaded gas and 2.The catalytic converter. My Hyundai's Owner's Manual describes their converter as follows: a one or two monolith type three way catalytic converter to

reduce the carbon monoxide, hydrocarbons and nitrogen oxides contained in the exhaust gas of the car. This meets the requirement of the various Clean Air Acts 1970, 1980 & 1990. Note well: it doesn't eliminate these gases, only reduces them. Also, it doesn't do anything to deal with carbon dioxide, the major cause of the greenhouse effect. In the next chapter, we will go over the various exhaust gases that come out of cars.
(1) Before doing this I would like to give you a brief overview of an internet fact sheet provided by the U. S. Environmental Protection Agency, Office of Mobile Sources. This should enlighten you concerning the dangers of the pollutants now poisoning our air, and clear up any doubt that the Clean Air Acts have done any serious curtailment in eliminating these poisons.

    1. Pollution from Cars – the average person's car is the single most polluter, since emissions from millions of vehicles on the road add up. Operating a private car is probably the average citizen's most polluting daily activity.

    2. Auto Emissions Sources – Car pollution comes from by-products of the automobile's combustion process (exhaust) and from

evaporation of the fuel itself. This comes about in three ways: Evaporative Fumes from the engine, Refueling Losses and Exhaust Emissions.

3. Combustion Process – the fuel in a vehicle is a mixture of hydrocarbons; compounds which contain hydrogen and carbon atoms. If an engine works perfectly, oxygen in the air would convert all the hydrogen in the fuel to water and all the carbon in it to carbon dioxide. Nitrogen in the air wouldn't be affected. However, in the real world the combustion process cannot be perfect, and car engines will emit several types of pollutants.

4. The major exhaust pollutants are: Hydrocarbons, Nitrogen Oxides, Carbon Monoxide & Carbon Dioxide. All of them will be discussed in the following chapter.

5. Evaporative Emissions – Hydrocarbon pollutants make it into the air via fuel evaporation. Since modern exhaust emission controls and improved gasoline formations lower hydrocarbon emissions from the tailpipe, evaporative loses can account for most of the total hydrocarbon pollution from current model automobiles. These emissions occur especially on hot days, when ozone levels

are at their highest.   The following is a list of various ways evaporative emissions happen:

One, Diurnal – as the temperature rises during the day, heating the fuel tank and venting gasoline vapors.

Two, Running Losses – the heat from the engine running can vaporize gasoline.

Three, Hot Soak – even when the engine is turned off, it remains hot for a period of time, giving off gasoline vapors.

Four, Refueling – The fuel tanks always have gasoline vapors present. These vapors escape, when the tank is open and filled with liquid fuel.

Nationwide: What has been accomplished to deal with Car Emission?

The Clean Air Act of 1970 – gave the EPA broad authority to regulate motor vehicle pollution.

Fundamental improvements in engine design, plus the addition of charcoal canisters to collect hydrocarbon vapors and exhaust gas recirculation valves to reduce nitrogen oxides.

Catalytic converters in 1975 reduced hydrocarbon and carbon monoxide emissions. Since lead inactivates the catalyst, unleaded gasoline was introduced into the pumps. This brought about a reduction in lead levels, alleviating many environmental and human health concerns.

In 1980 the three way catalytic converter was introduced. This converts carbon monoxide and hydrocarbons to carbon dioxide and water. It also helps reduce nitrogen oxides to elemental nitrogen and oxygen.

What has emission control done to improve Air Quality?

Since 1970 government with the cooperation of industry has reduced typical vehicle emissions. With that said, what they did, did not feed the bulldog. He's still hungry. To put it more to the point: this didn't solve the pollution problem; it still exists. In the same time frame Americans own more cars and drive more miles: double the number of cars and double the amount of miles traveled. This has off set much of the emission control progress. The net result is a modest reduction in automobile pollution except for lead.

However, ozone continues to be a persistent urban air pollution problem. Secondly, carbon monoxide remains critical in many large cities. Third, vehicle generated carbon dioxide is essentially unchecked by the Clean Air Laws. Being a major contributing greenhouse gas, the element's control must be given serious attention.

(2) From another internet source of information from the EPA, entitled: Automobiles and Ozone. The EPA speculates: the continued growth in the number of vehicles and miles traveled will result in an increase of hydrocarbons emissions from conventional gasoline vehicles, despite continued improvement in emission control systems. The EPA suggests an introduction of vehicles designed specifically for optimal performance on clean fuel. The following is a list of these fuels:

Alcohols: Methanol and Ethanol – cars designed to run on these fuels have the potential to emit 80 to 90 percent less reactive hydrocarbons.

Electricity: Battery Powered Cars – Although technology is limited, promising developments may lead to more widespread use in the future.

Natural Gas: is an excellent automotive

fuel, particularly for fleet vehicles where long driving range is not important. They have the potential to emit 85 to 95 percent less reactive hydrocarbons.

Liquid Petroleum Gas (Propane): This is a byproduct of petroleum and natural gas production, and emits less ozone forming hydrocarbons.

Reformulated Gasoline: This is a study by the petroleum industry to reduce hydrocarbon emissions. It is suggested by the EPA that a reduction of 15% will be required in high ozone areas to quell future pollution gains.

Source: (1) U. S. Environment Protection Agency, Office of Mobile Sources, EPA 400-F-92-007, Fact Sheet OMS-5 August, 1994, Automobile Emission: An Overview, Pages 1 to 4.

(2) Automobiles and Ozone EPA 400-F-92-006, Fact Sheet OMS-4, January, 1993, Pages 1 to 6.

CHAPTER SEVEN

MAJOR AIR POLLUTANTS

1. (1) Ozone gas can be found in two places: close to the ground (the troposphere), becoming a major part of smog (bad), and higher in the air (the stratosphere), where it helps block the sun's radiation (good). Ozone is created, when nitrogen oxide compounds mix in sunlight. Nitrogen oxide is a by-product of burning gasoline, coal or other fossil fuels. Ozone close to the earth is responsible for a number of health problems, such as: frequent asthma attacks in people, who have this disorder; also it can cause sore throats, coughs, and breathing difficulties, and in some cases lead to premature death. The gas can also harm plants and crops.

2. (1) Carbon monoxide gas comes from the burning of fossil fuels, mostly in cars. It cannot be seen or smelled, and is released when engines burn oil related fuels. Cars are responsible for most of the gas found outdoors. Also the emissions can be harmful in the home, if the furnaces and heaters aren't properly maintained.

From a physiological point of view, Carbon monoxide makes it difficult for the body parts to get the oxygen needed to run correctly. The immediate effect of the gas's exposure makes people dizzy and tired and gives them headaches. Older people with heart disease should especially be aware that high exposure to the gas can result in hospitalization or death.

3. (1) Nitrogen Dioxide is a reddish-brown, strong smelling gas (at high levels), which comes from the burning of fossil fuels from power plants and cars. Nitrogen dioxide is formed in two ways: when nitrogen in the fuel is burned, or when nitrogen in the air reacts with oxygen at very high temperatures. The gas can also react in the atmosphere to form ozone, acid rain and particles. (Acid rain can harm plants and animals, and render lakes dangerous for swimming and fishing). Exposure to the Nitrogen Dioxide, especially at high levels, can inflict coughs and shortness of breath in humans. Those exposed to it for a long time have a higher chance of getting respiratory infections.

4. (1) Greenhouse gases (carbon dioxide, methane and nitrous oxide) stay in the air for a

long time, and result in warming up the planet by trapping sunlight. This is called the "greenhouse effect,' because the gasses act like the glass in a greenhouse. (2) Of the three, Carbon Dioxide is the most distressing greenhouse gas; since it comes from the largest source: the burning of fossil fuels in cars, power plants, houses and industry. It makes up 83.7% of the total emissions. Methane is also released during the processing of fossil fuels, but in lesser amounts (11.4% of the total); it also comes from natural sources like decaying vegetation. Nitrous oxide (2.6% of the total) comes from industrial sources and rotting plants.

The concern of the greenhouse effect is that it might lead to changes in the planet's climate. This has been corroborated in recent studies. Their findings are: the hottest year on record, based on global average occurred in the 1990s. Over the past century, the Earth's average temperature has risen by approximately 1 degree F. All things being equal, scientists believe the next century could increase that amount by 2 to 6 degrees F. Some of these climate changes might include more temperature extremes, higher sea levels, variations in forest composition and crop production, damage to land near the coast, and natural habitat. Human health could also be affected by diseases that are related to temperature or by damage to land and water.

The major Greenhouse polluters are: United States followed by the former Soviet Union, China, Japan, Germany and India.

Source: (1)Time Almanac 2003, Page 592, Their source: Jonathan Levy, Harvard School of Public Health.

(2)U. S. Environmental Protection Agency

Note: We are so concerned with the greenhouse pollution, my partner wrote a separate chapter on it, entitled Global Warming. This is coming up next.

CHAPTER EIGHT
GLOBAL WARMING

Is global warming a problem? After all the average temperature change has only been an increase of one degree F in the past century. How can this be a problem? Doesn't seem to be much of a change; does it? We know the ice caps are melting, and if they are melting then it holds other ices in the world are melting too, creating more water. Should mankind be concerned about more water? I am. Two issues concern me: one, climate change and two, pestilence.

Climate change – there doesn't appear to be too much change as far as we all can see: the four seasons come and go as they always have, hot, cold, snow, rain. If it is hotter, I don't notice it, and nobody else notices it either. So what is the problem?

The problem may be just that – you don't notice it. If people don't see a problem, they don't react to it. The problem is a gradual one; like Cancer (the silent killer), it is there in the body eating away, until it is too late to do anything about it. The same holds true with pestilence: bugs and germs are still doing the same annoying thing they always did – colds, flu, mosquitoes, flies. There doesn't seem to be any difference. But is there a difference? I believe there is, and will be.

Let's take a look at climate first. The Red River (of the north), is a navigable river of the United States and Canada, arising in Elbow Lake, Minnesota, near the source of the Mississippi River, and flowing south and west to Breckenridge then north forming the boundary between Minnesota and North Dakota. As rivers go, it is rather insignificant. What called it to my attention is the fact that it doesn't usually flood. However, it has on April, 1997 and recently on April, 2009 driving residents of Grand Forks, ND and East Grand Forks, MN from their homes, many of which were destroyed. The reason given was the melting of ice, and we all know what melts ice: heat. By the numbers this catastrophe is a relatively minor one. However, it isn't isolated, in fact it is rather typical. I could cite many more similar climate changes, especially in the Mississippi River area, splitting the United States in half. Even though my sympathy goes out to the many people, who suffered through these events, I don't want to fill up several pages with this information. My point is, flooding is common today, and unusual in that it is of historical proportions.

(1)To give you a super catastrophe, it would be Cyclone Nargis, which affected: Bangladesh, Burma, India and Sri Lanka. Nargis was a strong cyclone that caused the worst natural disaster in the recorded history of Burma (officially known as Myanmar). The storm made landfall in the country on May 2, 2008, causing catastrophic destruction, where 146,000 fatalities were reported with thousands more people still missing. Damage was estimated at over $10 billion dollars, making it the most damaging cyclone ever recorded in this basin.

Nargis is the deadliest named cyclone in the North Indian Ocean Basin, as well as the second deadliest named cyclone of all time, behind Typhoon Nina of 1975. There were other unnamed storms: 1970 Bhola cyclone, the Bangladesh cyclone of 1991 and the Cyclone Mala of 2006. All of them were of serious proportions and casualties.

(2)   I'll give you one more climate change storm, which should make my point.   This is a catastrophe, closer to home.   On August 29, 2005, Hurricane Katrina lashed the Gulf coast (New Orleans, LA mainly) devastating the area 100 miles from the center of the storm.   1,836 people lost their lives during it, and in subsequent floods.   The hurricane was the costliest and one of the deadliest in the history of the United States.   Americans aren't used to storms of this nature; they took it hard.   But did they learn anything from it?

Up to this point, I gave you the wet climate changes of note.   It doesn't end there; there were also dry climate changes of significance:

February 6, 2009, China suffered the worst drought in 50 years.   The land is parched and the irrigation dams have dried up; crops and livestock are dying.

January 30, 2009, Melbourne Australia counts heat wave deaths.

September 26, Argentina grapples with fierce drought.

July 24, 2007, Hungary, deadly heat wave grips the country.

June 22, 2006, Washington DC study – Earth 'likely' the hottest in 2000 years. Were global temperatures recorded back then?

August 14, 2003, France, 10,000 dead in heat wave. Paris morgues are full and air conditioned tents have been set up to hold bodies. Officials blame the high death toll on the length of the heat wave and that Parisian buildings were typically lacking air conditioning.

May 9, 2000, India. Drought

Getting back to 'my point'. My point is, if all of the above climate changes were caused by one degree F increase in the Earth's average temperature, what changes will occur in a 2 to 6 degrees increase, predicted by scientists in the next century? That is all things being equal, and the world continues on in the same path of destruction, it is on now!

(1) Source: Cyclone Nargis – Wikipedia, the free encyclopedia, Wikipedia is a registered trademark of the Wikimedia foundation, Inc. a US registered tax-deductible nonprofit charity. Page 1

(2) Source: World Disasters. Weather Related Timetable, 21st Century, Copyright CNT Group, 2000

Next is pestilence. If there is more water and heat from the greenhouse effect, what does this suggest? Logically it tells me, there will be an increase in bugs, germs and viruses, which survive and incubate well under these conditions. Three incurable diseases worry me the most; there are many more, but I want to concentrate on these three, since they are the most threatening:

(3) AIDS OR ACQUIRED IMMUNE DEFICIENCY SYNDROME:

The following is information about this fast spreading contagious disease:

1. It is characterized by the failure of the immune system, making those affected more likely to develop infections such as pneumonia, tuberculosis and cancer. These victims have a 90% chance of dying from one of those side diseases.

2. Transmitted sexually and via intravenous drug users.

3. The worse area infected is the Sub-Saharan Africa, 28.1 million, but the illness is rapidly spreading to every corner of the world.

4. The human immunodeficiency virus (HIV)

that causes AIDS was identified in 1983, and by 1985 tests to detect the virus were available.

MALARIA:

For some reason this disease and Tuberculosis are considered dead illnesses. Believe me they are not. Malaria is the second largest killer (after tuberculosis). Half a billion people suffer periodic attacks of the disease, it kills an estimated 2 million people each year. Approximately 40% of the world's population lives in high-risk areas, which are categorized by rural living without clean drinking water and adequate health facilities.

TUBERCULOSIS (TB):

TB is responsible for 3 million deaths each year. About one third of the world's population is infected by it. An added danger is the apparent alliance between TB bacillus and HIV. The latter renders a TB carrier 30 times more likely to develop active TB. The World Health Organization predicts by 2020 nearly 1 billion people will be newly infected with TB.

(3)Source: Natural Hazards - Pestilence & Disease, Page 1, NDA Natural Disaster Association, Information Resource on Natural

Hazards & Disasters, March 28, 2009

CHAPTER NINE

(1) RADIOACTIVE STRONTIUM AND BONE CANCER

Note: The following is the research from various noted research scientists concerning what the use of Nuclear Energy is doing to humanity. If you want to know why leukemia (and other common diseases) are on the rise, then this will give you an idea.

(2) Dr. A. V. Topchiev, an academician from the former Soviet Union, was one of the first researchers to discover the dangers of radioactive fallout. Essentially, he revealed that the world already has dangerous amounts of radioactive strontium in the air, which has penetrated our bodies; specifically our bone structure. This is the result of nuclear bomb explosions and nuclear testing of all kinds. Dr. Topchiev's warning is within the next decade radioactive strontium poisoning will reach serious levels, and sarcoma of the bone and leukemia cases will become a major concern.

Strontium 90 is the radioactive form of strontium, and is man-made through nuclear fission. It is a chemical cousin of calcium, which makes it a great danger to us all, since it acts like and, has similar characteristics to calcium. Strontium 90 is often absorbed by plants and animals, when there isn't enough calcium. Since all living things need calcium, their draw to strontium 90 contamination becomes a logical progression. The fallout from it is now found in every corner of the earth, in every continent and is especially prevalent in milk and cereals. With each new nuclear test or accident, the strontium 90 increases accordingly everywhere in the world.

Strontium 90 and its lesser radioactive elements like cesium 137 (a chemical cousin of potassium, and also a product of fallout) remain in a human body throughout a lifetime. The radiation from it produces bone cancer, anemia and leukemia (cancer of the blood).

(3)Yukio Tanaka and Stanley C. Skoryna, two noted research scientists, McGill University, Montreal, Canada have similar views to Dr. Topchiev. They describe Strontium 90, when ingested, as a bone-seeking element absorbed through the intestinal wall and then deposited in the bone. As a sufficient amount of it is accumulated in the bone, bone tumors and other malignant changes occur.

Bone cancer strikes mostly the young. This is the reason given: radioactive strontium, as an elemental pollutant, falls from the air onto vegetable matter, where it is absorbed by plants with a naturally high content of calcium. Grass, one such plant, is eaten by dairy cows; hence radioactive strontium gets into their milk. Since children drink larger quantities of milk than adults, they are more prone to the disease.

How does the radioactive strontium get into the air, now that above-ground tests of nuclear bombs have long since been discontinued?

(3) Professor Ernest J. Sternglass, University of Pittsburgh, Radiology Department in his studies has shown that the installations processing nuclear fuel for power plants, and the power plants themselves, release radioactive gases into the atmosphere and thus increase fallout around each plant.

Additionally, as each nuclear plant gets older, the quantities of radioactive pollutants released, increases each year.

Source: (1) Encyclopedia of Common

Diseases, by the Staff of Prevention Magazine, 1976 by Rodale Press, Inc., Page 281

(2) Intestinal Absorption of Metal Ions, Trace Element and Radio Nuclides, (Pergamon Press, New York, NY, 1970) by Yukio Tanaka and Stanley C. Skoryna

(3) Low-Level Radiation (Ballantine Books, New York, NY, 1972) by Professor Ernest J. Sternglass of the Department of Radiology, University of Pittsburgh

CHAPTER TEN
CANCER

Cancer research has determined that Cancer isn't just one disease; it is a group of many different diseases i.e. lung cancer, breast cancer, colon cancer, leukemia, bone cancer, etc. What groups them together is one important characteristic in common: they affect the body's cells, the basic unit of life.

In the cancer progression, normal cells for some reason become cancerous or abnormal. In the normal process, cells grow, divide, and produce more cells, needed to keep the body healthy. However, the process sometimes goes astray: cells keep dividing, when new cells are not needed. The mass of extra cells forms a growth or tumor, which could be benign or malignant.

Benign tumors are not cancerous, and can be removed without the concern of coming back. Also they do not spread to other parts of the body, and are rarely a threat to life. Malignant tumors, on the other hand, are cancerous. Cells in these tumors are abnormal, and divide without control or order. In addition they can invade and destroy the tissues around them. What is most disturbing about cancerous cells is, they can break away from a malignant tumor and enter the blood stream or lymphatic system (these two networks of vessels carry blood and lymph throughout the body), and spread to other parts of the body. This process is called metastasis.

Source: National Cancer Institute
Time Almanac 2003 with Information Please

LUNG CANCER FROM THE AIR WE BREATH

Since 1914, Lung Cancer among males has gone up nearly 2,000%. A noted researcher, Dr. Eugene Handry, who spent a lifetime studying petroleum chemistry, reported that this increase corresponds exactly with the increase of gasoline consumption. Also noted in his work, lung cancer declined 35% during the war years, 1941–1945, when gas consumption was reduced because of rationing. It is not difficult to understand this in light of the fact that 600,000 tons of pollutants are poured into our atmosphere daily. Besides the known poisonous gases such as carbon monoxide, nitrogen dioxide and nitrogen oxide (all tail pipe gases), there are also particles of lead and a number of hydrocarbons which are carcinogenic. These carcinogens make up a large percentage of the pollutants and are highly virulent.

Source: Encyclopedia of Common Diseases

Note: In light of the above, shouldn't there be a warning statement placed on all automobiles, clearly stating the following: "the breathing of auto exhaust can be hazardous to human health." The sign would be a constant reminder to the public, what their beloved cars are doing to them, their children and the earth.

Why doesn't Congress pass a law to this effect?   Do their constituents ever mention it to them?   Why don't you find out: ask your Congressman.   I'll bet he won't agree with you, and if he does, he still won't do anything about it anyway.   My fear here is that the oil/automobile industries are more powerful than Congress. America is in deep trouble, if this is true!

(1)   EXERCISE AND CANCER

There have been some researchers, who suggest that one of the best preventive measures against cancer may be exercise.   The work of (2)

R. A. Holman, M. D. Honorary Consultant Bacteriologist, United Cardiff Hospitals and (3)Dr. O. H. Warburg suggests, impaired utilization of oxygen by the body's cells may be responsible for the beginnings of many cancerous growths. Strong breathing from good exercise is one of the best ways to insure the entire body is receiving enough oxygen.

The importance of the oxygenation process and cancer prevention appears to have been borne out in a study conducted in Germany by Dr. Ernst Van Aken in 1971. He evaluated the medical histories of runners belonging to a long-distance running club; they consisted of 454 members of older long-distance runners. He found that in the six-year period of his study, only four of the runners got cancer, and all four recovered and are now running again. To prove his findings, he also kept records of a parallel group of 454 men in the same age bracket, who didn't run. 29 cases of cancer during the same period were recorded with seventeen of the no-runners dying of the disease.

Dr. van Aaken's findings were reported in the January 15, 1971, article in the Cologne Stadt-Anzeiger. He emphasized in the report that the low-Cancer group formed no human elite, but was simply a group of older men, who ran three to five miles a day. He reasoned that anybody in reasonable health can get the same results – if they motivate themselves to get out and exercise every day. Other "huff-and-puff" exercises like cycling, climbing, long-distance swimming and cross-country skiing are just as good. Dr. Van Aaken concluded, the reason the runners didn't get cancer is that they were constantly providing their bodies with more oxygen than needed.

Source: (1) Encyclopedia of Common Diseases
(2) The Cancer Problem and How to Cope with It, by R. A. Holman, M. D., Honorary Consultant Bacteriologist, United Cardiff Hospitals
(3) New Methods of Cell Physiology, by Dr. O. H. Warburg

Note: I know this isn't a total cure for the disease, but why not do it anyway. It's better than doing nothing at all. Do something: run, walk, lift weights, jump up and down, any action, which will put your body in gear. The old saying applies: "if you don't use it, you'll lose it."

CHAPTER ELEVEN
OZONE HOLE

Ozone is crucial or harmful for life on Earth, depending on where the gas resides. The good placement of Ozone is the layer found in the stratosphere about 7 – 28 miles high. It shields the Earth's surface from the sun's damaging ultraviolet rays. The bad Ozone is within breathing distance near the Earth, as was previously discussed under Major Air Pollutants.

Briefly, Ultraviolet Radiation consists of electromagnetic waves between the violet and the visible band and X-rays. It is present in sunlight, and has an important role as a photochemical agent in certain life processes i.e. florescence, vitamin D and Spectroscopy. In other words, within a certain amount of the rays, it is good for humans, animals and plants. However, too much of a good thing isn't always good. Too much exposure to ultraviolet rays on Earth could be disastrous. It can lead to skin cancer and eye problems, and also harm plants and animals.

A disturbing finding occurred on September 2000 by NASA scientists, who detected a large hole in the Ozone protection; around three times the area size of the United States. The hole is located over the Antarctic with total ozone of 200 Dobson Units or lower. The data was acquired by the Ozone Monitoring Instrument on NASA's Aura satellite.

To qualify what an Ozone Hole is, it actually isn't a hole where no ozone is present, but is a region of exceptionally depleted ozone in the stratosphere. This wasn't observed prior to 1979. The cause for alarm is how much of this depletion is going to happen, until it is corrected and returned to natural levels.

Paul Newman, a research scientist at NASA's Goddard Space Flight Center had this to say about the problem. Over areas farther from the poles, Africa or the United States, the ozone levels are only three to six percent below natural levels; over Antarctica, the ozone is 70% lower. NASA has an improved computer model, which allows their scientists to more accurately estimate the ozone depletion over Antarctica, and how the hole will reduce in time.

Newman's findings aren't as optimistic as before; he and his colleagues figure the hole will not start shrinking significantly until 2018, after which time the recovery should proceed more quickly. All things being equal, the hole won't replenish until 2065.

In conjunction with the NASA estimation on the ozone hole replenishment, thought must be turned to what is causing the problem. Does NASA know? Here is what they are basing their study on: the hole is caused by – chlorofluorocarbons (CFCs), halons and other compounds that include chlorine or bromine. CFCs are used in air conditioners and refrigerators, since they work well as coolants. They can also be found in aerosol cans and fire extinguishers. Other stratospheric ozone depleting substances are used as solvents in industry. Efforts to curb those chemicals have in recent years led to optimism that ozone would rebuild.

But is that information entirely correct? My partner and I have our doubts, but we can't prove it, so we don't want to say that fossil fuel's emissions have something to do with ozone depletion, but the suspicions persist.

Source: NASA Ozone Hole Watch: What is the Ozone Hole?
October 4, 2004, NASA Official: Paul Newman

CHAPTER TWELVE
THE MARS DISASTER

One thing was lacking with my piece on NASA's ozone hole discovery: I neglected to cross the plain concerning why the phenomena is so important to mankind. Discussing the subject is like saying the unspeakable. What will be the consequences and the disasters that will befall the Earth without the ozone protection from the sun's ultraviolet rays? As previously gone over in detail, Ozone placement is one of those delicate balances, which makes life function on the planet; take away its' location surrounding our globe, and life as we know it will cease to exist.

What is intriguing about our neighboring planet Mars, and why I feel it belongs in an environmental book about earth, is the remarkable similarities between the two planets. There is a great deal that can be learned about Earth by studying Mars. From the recent space craft landings on the red planet there is strong evidence that life on that planet may have existed many years ago. Yet, photos of the red planet show it uninhabitable: a variable desert of rock and sand, void of any of the ingredients which support life i.e. surface water, oxygen, shelter from the sun, etc. Simply put the neighboring planet in its' present state is useless to mankind as a place to live. Moving along logically on that note, if life did exist on the planet's surface, why did it cease to exist? Could it be the planet lost its' ozone protection?

To further evaluate this possibility, we want to first list data concerning the red planet, as it now functions in our solar system. This will give you an idea why many scientists consider Mars as Earth's sister planet. From there we will devote most of the chapter reviewing the incredible information gathered from the recent space craft landings on the planet. We feel all this drives our theory: that the Mars' failure as a living planet came about because of their loss of ozone protection.

1. Mars axis rotation is nearly the same time as Earth – 24 hours, 37 minutes making the Mars day almost identical to Earth.

2. A Mars' year takes 687 days to circle the Sun. The reason for this is Mars has a more oval path around the sun, and can vary by as much as 200 million miles from Earth's.

3. Mars atmosphere is much thinner; atmospheric pressure is about 1% that of our planet.

4. Being a smaller planet its' gravity is one third of Earth's.

5. Major constituents are carbon dioxide and nitrogen. Water vapor and oxygen are minor constituents. Mars' polar caps are composed mainly of frozen carbon dioxide (dry ice); they recede and advance according to the Martian seasons.

6. Like Earth, Mars has four seasons, but they are much longer. In their northern hemisphere, the Martian spring is 198 days, and the winter season lasts 158 days.

The Pathfinder lander and its rover, Sojourner, landed on Mars, July 4, 1997, providing scientists with a wealth of information concerning the surface and atmosphere of the planet.
The lander returned live pictures of the planet's topography, and the rover explored the mineral composition of a variety of rocks on the surface, with its cameras and on-board X-ray spectrometer. The mission in three months of operation returned more than 16,000 images of the

Martian landscape from the lander's camera and 550 images from the rover.

It was discovered in the study, the reddish surface soil was due to the presence of oxidized iron (rust) in it. Sojourner soil samples taken from several sites revealed their composition was similar to those analyzed by the two Viking landers in 1976. This might be a consequence of the Martian winds distributing the soil evenly over the planet.

The returned data surprisingly revealed how rapidly the Martian temperature fluctuated (30 to 40 degrees in a matter of minutes); possibly due to atmospheric turbulence from strong, gusty winds carrying warm air from one region or cold air to another. Temperatures range from 80 degrees F at the equator during the day to minus 199 degrees F at the poles at night.

The Pathfinder provided pictures and subsequent data, which gave strong evidence, Mars had an abundance of water on their surface millions of years ago. Scientists have inferred from the variety of rocks and sediments found in the Ares basin (landing sight) that the channel was once awash with torrential floods greater than any on Earth. The diversity of rocks deposited there suggests, they were moved down from the highlands from the surface flooding.

NASA in 2000 announced information from the Mars Global Surveyor (part of the Pathfinder mission) indicating features that looked like gullies carved out by flowing water and deposits of soil and rocks transported by the flow. Since gullies had never been reported on Mars, the new information gave rise to the suspicion, there might be current sources of liquid water at or near the surface.

Two years later, NASA had received information from their Mars' Odyssey Space Craft, launched April 2001, that their water suspicion was correct. The craft had found large quantities of water below the surface of the planet.

Source: Time Almanac 2003, Copyright 2002 by Family Education Network, Inc., Time Inc. Home Entertainment, Page 383

Authors' Comments: So the scientists' water theories are looking more and more correct, especially in light of NASA's Mars' Odyssey spacecraft findings and the Pathfinder/Sojourner's many photos and rock samples of the red planet

that were evaluated. If water, then life would be the logical progression. And if this is a truism, then what happened to it on Mars? Using logic as a tool, there must have been something catastrophic, which made its' demise occur. What was it? My guess is a condition took place, which made the planet lose its' Ozone ultraviolet rays' protection. It probably began as a hole in the Ozone gases surrounding the planet and escalated from there. It was a slow process, perhaps, taking several centuries to come to fruition. It was steady and nothing was done to reverse its' progression. Intelligent beings, if they existed, died off from lack of food and water and from physical ailments, like cancer and blindness.
It was painful for all the creatures, who depended on resources from the planet. There was nowhere to hide or guard themselves from the conditions that were unraveling. Death and destruction were eminent. Now all that is left on the planet's surface are rocks and red clay, and a couple of robotic machines taking pictures of a barren piece of rock circling the sun going nowhere, doing nothing. A monument saying, "Don't interfere with the delicate balances of nature." Will we respect

this and listen to our scientists, who are presenting convincing data of this possibility to our sister planet, or will we ignore them? Yes I know it is only a theory, but it is a good one backed up by facts, photos, and scientific evaluation. The warnings are clear, stop destroying the earth's ozone layer! If this isn't attended to then Pathfinder's pictures of the red planet will be what the earth will look like in the future. Allow me to repeat myself, when the catastrophe strikes, the erosion will be a slow painful death for every live creature on earth. At the rate we are going, it could begin in a matter of a mere several generations from now.

CHAPTER THIRTEEN
PHOTOSYNTHESIS

Photosynthesis is an important life process on the planet. It is a unique, metabolic process, occurring in green plants, which utilizes light (mostly sunlight) to convert carbon dioxide and water into carbohydrates and oxygen. Almost all plants and animals dependent upon photosynthesis for survival. It is the chief means by which energy from nonliving sources is transformed into chemical energy useable in the life processes. While it is true green plants undergo normal respiration: absorbing oxygen and excreting carbon dioxide; in photosynthesis they absorb about five times as much carbon dioxide as they excrete, and release about five times as much oxygen as they consume in respiration. The rule-of-thumb is 180 square inches of green-leaf surface during an average summer can supply the annual oxygen requirements for a human being. Terrestrial plants (worldwide) produce about 88,000,000,000 pounds of photosynthesized carbohydrate material per year.

Photosynthesis is a process in aquatic and

terrestrial plants alike: the major difference between the two is aquatic plants absorb their carbon dioxide from and excrete their oxygen into water, whereas terrestrial plants do so into the air. Different colors of sunlight or similar artificial light such as red, blue, indigo, and violet wave length influences the chloroplasts, or chlorophyll-containing cell bodies, which absorb carbon dioxide. The latter originally enters the plant through the stomata (q.v.), or pores, on the plant's external surface.

Source: Funk and Wagnall's Encyclopedia, Page 7059 –1963

Note: Photosynthesis, through light and plants plays an important role in the life cycle balance. The following are some facts about forest deforestation, prefaced with information concerning the Amazon River. (note: I'm adding the Amazon information, since it supports the great Brazilian forest area, which is now being deforested leading to less plants and equals less oxygen and carbohydrates.)

AMAZON RIVER

One of the largest, perhaps the largest, rain

forests is in the Amazon River area of Brazil. To give you an idea of the scope of it, I've put together some statistics about the river. It is so wide that from its mouth pours one-fifth of all the moving fresh water on earth. In addition to being the widest, the Amazon also may be the longest (estimates vary up to 4,200 miles), covering the largest area (2,772,000 square miles), and having the greatest discharge (7,200,000 cubic feet per second). The area drained by it is about the size of the mainland United States. 85% of it hasn't been explored.

Source: Isaac Asimov's Book of Facts, Bell Publishing Company, New York, NY, Copyrighted 1979 by Red Dembner Enterprises Corp., Page 128

RAIN FORESTS: DEFORESTATION

A rain forest is a hot wet equatorial forest characterized by high, broadleaved, evergreen trees and the absence of undergrowth.

Largest deforestation in history occurred in Brazil between 1990 and 2000; the country cleared an average of 8,596 miles of forests annually in that period to meet demands for land and timber.

Between 1994 and 1995, deforestation rates in Brazil nearly doubled to about 11,197 miles yearly; this is the largest increase in the deforestation rate ever. Since then, the rates have returned to the late 1980's and early 1990's levels – about 5,000 miles per year.

However, the fastest forest depletion wasn't in Brazil, but in Burundi, central Africa. Their forests were being depleted at an average of 9% every year between 1990 and 2000. Hopefully the Burundi government will bring this under control, or they won't have any forests at all. Estimated depletion will occur in 11 years at this rate.

Source: Guinness World Records 2003, Page 64

ACID RAIN – (more plant damage)
(See Nitrogen Dioxide on page 12 for description of Acid Rain)

Most acidic acid rain – The condition is categorized with pH 7 being a neutral reading (the lower the number, the higher the acid level), the lowest pH level ever recorded for acid precipitation is a reading of 1.87 at Inverpolly Forest, Scotland,

UK in 1983. This area has the distinction of receiving the most acidic acid rain.

The second category: Worst Acid Rain Environmental Damage –
The Czechoslovakia Republic wins this award, having the greatest forest damage due to acid rain with 71% of them affected. The problem is particularly acute in northern Bohemia, where the pollution is caused by large amounts of fossil fuel gases in the atmosphere, which eventually finds its' way to the earth and forests. The fallout has been traced as coming from Polish and East German industries.

Source: Guinness World Records 2003, Page 64

Note: When is man going to realize, his best friend isn't a dog; it is plants?
(My apology to dog lovers. Rest assured, I didn't mean that negatively.)

CHAPTER FOURTEEN

NUCLEAR ENERGY (how it all began)

First nuclear chain reaction (fission of uranium isotope U-235) was produced at Chicago University under the direction of physicists Arthur Compton and Enrico Fermi on 12/2/42.

Albert Einstein, who was the first to split the atom making the Compton/Fermi discovery possible, had this warning concerning the creation of the Atomic Bomb. "A single bomb of this type, carried by boat and exploded in a port, might very well destroy the whole port, together with some of the surrounding territory." His assessment arrived on President Franklin D. Roosevelt's desk in 1939. Six years later, on July 16. 1945, the first Atomic Bomb was detonated, during a test in the desert near Alamogordo, NM. On August 6, 1945, the U. S. dropped a similar bomb on Hiroshima, Japan killing more than 70,000 and leveling the city. A second bomb was exploded over Nagasaki, killing more than 40,000, forcing Japan's surrender and concluding the war in the Pacific. The Hiroshima/Nagasaki deaths didn't end with the initial strike; radiation from it would kill or maim tens of thousands more. Having helped create this horrific tool of destruction, J. Robert Oppenheimer, the physicist who led the Manhattan Project, which developed the Bomb, would go on record as opposing the creation of the more powerful Hydrogen Bomb. His feelings echoed Einstein's, who later lamented, "If only I had known, I should have become a watchmaker."

Source: 60 years Life Magazine, A 60th Anniversary Celebration 1936 – 1996, Published by Life Books, Time, Inc., New York, NY, 1996 – Page 96

U. S. president Harry S Truman authorized production of the Hydrogen Bomb 1/31/50, i.e. larger more devastating nuclear weapon.

First hydrogen device explosion 11/1/52 at Eniwetok Atoll in the South Pacific.

Atomic Energy Commission (A. E. C.), a civilian agency working under the US government, was established through the Atomic Energy Act of 1946. The agency's purpose was to administer and regulate the country's atomic-energy activities. The revised Atomic Energy Act of 1954 would end government control of atomic-energy facilities. The work of the AEC, up until then, was primarily devoted to developing and producing a nuclear-weapons stockpile. Most aspects of its programs were classified. The revised act encouraged industry to develop nuclear reactors for generating electricity and to produce radioisotopes for medical and other purposes.

The A. E. C. has increasingly emphasized peaceful uses for atomic energy, which includes basic research concerning the structure of matter; nuclear-powered rockets for deep-space flight; radioisotopes for industry, medicine, and agriculture; water-desalination plants; explosives for large-scale excavation, and the recovery of natural gas, oil and minerals; and food preservation by irradiation. The A. E. C. under the Atoms-For-Peace program cooperates with other countries in developing their atomic-energy programs.

Source: Funk & Wagnall's Encyclopedia, Page 432

CHAPTER FIFTEEN
NUCLEAR WASTE (RADIOACTIVE)

(Note: See page 11 Radioactive Strontium and Bone Cancer.)

The disposal of nuclear waste has become a worldwide problem. The international nuclear industry will be faced with a need to establish a long term energy production plan, to include a standard safe method of waste disposal. The concern is that the current state of scientific knowledge is limited in predicting the extent that waste could find its' way from the deep burial back to contaminate drinking water. This was the central theme presented in a document from the International Atomic Energy Agency (IAEA), published 10/07/08. Essentially it says, there isn't a universal model for the dissolution of waste matter in relation to the disposal site. Although much research has been given to the matter, it is far from being finished.

The Department of Energy or D. O. E. (USA) admits there are, millions of gallons of radioactive waste as well as thousands of tons of spent nuclear fuel and materials and also huge quantities of contaminated soil and water (in the country). The D.O.E. in spite of these large quantities, has a goal of cleaning all contaminated sites by 2025. To give you an idea of what they have on their plate: the Fernald, OH storage site had 31 million pounds of uranium product, 2.5 billion pounds of waste, 2.75 million cubic yards of contaminated soil and debris. Secondly, a 223 acre portion of the underlying Great Miami Aquifer had uranium levels above drinking standards. The United States currently has at least 108 sites, designated contaminated and unusable; some locations have many thousands of acres. The D.O.A.'s goal is to clean or mitigate many or all locations by 2025; however they admit the task is formidable, and some sites will never be completely remediate. So in the same breath, they are saying they won't be able to accomplish their goal. Odd way of putting it; wouldn't you say? The Oak Ridge National Laboratory, one of the larger sites, had at least 167 known contaminated release tracts in their 37,000 acre location. In the smaller U. S. sites the D.O.E. has successfully completed cleanup, or at least closure of several of them.

Nuclear waste requires sophisticated treatment and management to successfully isolate it from interacting with the biosphere. This usually necessitates a long term management strategy involving storage, disposal or transformation of the waste into a non-toxic form. World-wide governments are considering a range of waste management and disposal options, which is fine. However, there has been limited progress toward a long-term standard waste management solution.

Note: Let me interrupt here: "...there has been limited progress towards long-term waste management solutions." Let us concentrate on that statement. When are they going to get going? Why can't they find a solution? It is about time they did!

Continuing on with the research data:

Disposing of High Level Wastes or H. L. W. – Final disposal of high level wastes is required in due course but there are no technical or logistical reasons why this is urgent; the longer HLW is in storage, the easier it is to handle safely. (This statement sounds stupid to me in light the next statement.) HLW is accumulating at about

12,000 tonnes a year worldwide. High-level wastes are highly radioactive for a long time, so this must be isolated from people for thousands of years, while their radiation levels drop. Sounds urgent to me!

Now getting down to the nuts & bolts or suggested disposal, called the multiple barrier concept:

... The waste, either as a ceramic oxide (e.g. the spent fuel itself) or through vitrification (separated HLW from processing) is immobilized.

... It is then sealed in a corrosion resistant canister such as stainless steel or copper.

... Finally it is buried in a solid rock formation. (Sounds like a standard to me; what did I miss in all this?)

Source: Radioactive Wastes: WNA 11/1/08, Copyright World Nuclear Association, All rights reserved. 'Promoting the peaceful worldwide use of nuclear power as a sustainable energy source.' Pages 1 to 7

Note: Common sense will tell you all this radioactive dumping is bound to catch up with the environment sometime in the future. They can't keep on dumping and dumping forever; something has to give. My question is, is there any research being done on how to constructively use the waste? Or is it just written off as non-profitable to invest in this sort of thing?

CHAPTER SIXTEEN
VESTED INTEREST

(1) George Walker Bush, 43rd United States President appointed his Cabinet (2001); it was made up of more multimillionaires than any of his predecessors. Out of 16 full cabinet members, the Bush administration had 13 millionaires, seven owned assets worth more than $10 million. The members acquired the nickname 'tycoon's club'. The wealthiest of the group were Defense Secretary Donald Rumsfeld and Treasury Secretary Paul O'Neil each having declared assets of at least $61 million.

Note: I wonder how many of these men had oil industry connections. I guess it isn't important. I do know that George W. Bush and his Vice President Richard B. Cheney did. Look at where the price of oil from the pump went during their terms in office. With all the money the oil industry made, what was done by them to clean up the exhaust from their product. All they seem to be interested in is drilling for more oil domestically.

2. To give you an idea of how much oil Saudi Arabia has: their country possesses the largest oil

field. The Ghawar field was developed by Aramco (Saudi Arabia), it is 150 miles long and 22 miles wide. Recently the oil field was estimated to contain between 70-85 billion barrels of proven reserves.

I'll bet these boys are looking forward to selling all of it at their prices.

3. Deepest Underground Nuclear Explosion occurred on June 18, 1985. The 2.5 kiloton nuclear device was detonated at a shaft 9,359 feet deep at a site 37 miles south of Nefte-yugamsk, Siberia, in the former USSR. I can't exactly say why this bothers me. I guess it has to do with the power of it and the unknown understanding of what radioactive fallout is, i.e. where it can go and what it can affect, even at that depth.

4. The United States is the largest consumer of fossil fuels and commercial energy (fossil fuels plus hydro and nuclear power). In 1998 the nation consumed 1,937 million tons of oil equivalent (Mtoe) of fossil fuels and 2,147 Mtoe of commercial energy.

5. Largest Man-Made Excavation – The Bingham Canyon Copper Mine near Salt Lake City, Utah, USA, is the largest man made excavation.

It measures 2.5 miles across and .5 miles deep and is visible from outer space. The work began on it in 1906. Since then more than 5.9 billion tons of rock have been excavated; it has produced over 15.95 million tons of copper, 1.6 million troy lbs. of gold and 15.8 million troy lbs. of silver.

Note: Can't help but wonder, when will they fill this hole up after it is depleted of its' minerals, or will they just leave it there, and what will this do to the environment of the area?

6. The worst marine pollution happened in a fertilizer factory on Minamata Bay, Kyushu, Japan. The factory (whether on purpose of accidentally) continued to deposit mercury waste into the sea between 1953 and 1967. Approximately 20,000 people were affected by the mishap. The disease from it affected 4,500 seriously, with 43 people dead. An additional 800 deaths were attributed to Mercury poisoning from related sources, while 111 other people suffered permanent damage.

7. The United States has the highest emissions of Carbon Dioxide, one of the key gases responsible for the "greenhouse effect." In 2000, the country emitted 11,600 billion pounds. (5,800 million tonnes) of CO_2, a 3.1% increase

from the previous year.

8. Mexico City, Mexico is the most polluted city in the world, exceeding the WHO (World Health Organization) guidelines in sulfur dioxide, ozone, suspended particulate matter, and carbon monoxide. The city also has moderate-to-heavy lead and nitrogen dioxide pollution. The mountains surrounding the community have a lot to do with the problem, since they act as a trap holding the pollution in place.

9. Lake Karachay in the Chelyabinsk province of Russia is the world's most contaminated lake. It has accumulated 120 million curies of radioactivity and absorbed nearly 100 times more strontium 90 and cesium 137, than was released at Chernobyl in 1986. Just standing near the lake's shore a person would receive a radiation exposure rate of 600 roentgens an hour. To give you an idea of the exposure potency: it has 2000 times greater radiation than that emitted during a normal chest X-ray, and is strong enough to kill a person in an hour. The lake is near the nuclear facility Mayak Chemical Combine. The facility was shut down in 1990, but its' emitted radiation is still dispersed by the wind, or is absorbed by sand

particles and sediment at the bottom of the lake.

10. The most acidic water runs through the Richmond Mine at Iron Mountain California. When evaluated in 1990, it had a pH value of −3.6 (the lower the number, the higher the acid count). The source of the acid in the water was hot acid solutions dripping off colored stalactites in an abandoned copper and zinc mine.

. 11. The most powerful thermonuclear device tested had the power equivalent to approximately 57 megatons of TNT, which was detonated by the former USSR above the remote Artic island of Novaya Zemlya. The 56,000 lbs. (28-tonne) bomb was air-dropped on October 30, 1961. The shockwave circled the world three times, with the first circuit taking 36 hrs. 27 min. Some scientists estimated the power of this device at between 62 and 90 megatons.

12. (1)A 104-kiloton nuclear device was detonated at the Semipalatinsk Test Site, Kazakhstan, former Soviet Union, 583 feet beneath the dry bed of the Chagan River on January 15, 1965. It left a crater that was 1,338 feet wide with a maximum depth of 328 ft. A major lake naturally formed behind the 65–114 ft. upraised lip

of the crater.

13. The Russian type 65 torpedo is the world's largest, with a 25.5 in diameter, and can carry either a warhead of nearly 2,204 lbs. (one tonne) of conventional explosive, or a 15-kiloton nuclear warhead. This gives the device approximately the explosive power of the atomic bombs which destroyed Hiroshima and Nagasaki in 1945.

14. The most costly war in terms of human life was World War II (1939-1945), in which the total number of military and civilian fatalities of all countries is estimated to have been 56.4 million. To put this in perspective: the United Kingdom's population today is 59.5 million. The country that suffered the most in proportion to its population was Poland, 6 million killed, or 17.2 % of its population of 35.1 million.

15. Harvard University (USA) is the world's wealthiest university. In 2000 their endowment was $19 billion dollars; larger than the annual operating budget of 142 of the world's countries. The college's alumni includes seven graduates who went on to be president of the US and fourteen Nobel Prize winners.

Source: (1) Guinness World Records,

2002, Page 116

(2) All others – Guinness World Records, 2003, Page 62 – 66

Note to the wealthiest university: One would think with all your money and brain power, someone connected with the college would come up with a solution to the world's environmental problems. Surely if he or she did, the monopolies, heads of countries, and big wheels in general would listen to what was offered, and put the solutions into practice. Just think, solving such a problem would probably mean another award for the university's collection.

CHAPTER SEVENTEEN
SOLUTIONS

"There are no problems, only solutions". John Lennon (former Beatle and peace activist)

The following are some possible solutions to the world's environmental problems:

1. Conserving energy – The obvious are: using compact cars instead of SUVs, which are no more than a pick-up truck with a box on the back. This has become the American's vehicle of choice. It burns on the average twice as much gas as the compact.   Secondly, using public transportation. In Philadelphia, we have the elevated train, which can get you into center city in fifteen minutes from either end of its' total route.   Third, bicycling and walking is good for the heart, but dangerous, if there isn't a bicycle path or side walk.

2. Natural Gas – a combustible gas; issuing from the earth's crust through a natural or artificial opening, often associated with petroleum.   It is used extensively as a fuel gas, and sometimes as a source of other chemicals.   Gives off few dangerous emissions.

Lowest emissions car is fueled by natural gas. It was introduced in 1998, by Honda Motors (their Civic GX) is the world's most eco-friendly natural gas vehicle. Emissions of Carbon monoxide, Hydro carbons and Nitrous oxide from the car have been reduced to almost zero, while Carbon dioxide emissions have been reduced by about 20%. The Civic GX has 98% fewer emissions overall compared to a standard low emissions vehicle.

3. (1) The same company, Honda Motors, has developed a low emission petroleum vehicle; their Accord EX Sedan, which has the lowest emission levels of any gas-fueled car. Certified by the California Air Resources Board (CARB); the engine of this car emits only 2.3 lb of ozone-forming hydrocarbons during 100,000 miles of driving, about the same as spilling 32 fl. oz. of gas. This is an 86% reduction compared to the normal low emission vehicle in its class.

4. Japan's Kobe-Eco-Car car rental company has the lowest-emissions fleet of automobiles. It is dedicated to the rental of environmentally friendly cars, and has a fleet of 53, made up of electric vehicles, compressed natural gas vehicles,

and hybrid cars. They are new in the industry (established January, 1999), but definitely on the right track.

5. Between 1990 and 2000, China has organized the largest reforestation program. During this time they replanted enough trees to cover an average area of 6,974 miles every year (equivalent to the size of Kuwait). Loans from the World Bank support Chinese endeavors in this program.

6. Switzerland leads the world in recycling glass, with an estimated 91% of the glass products sold being recycled. The country's tradition of recycling and its network of free collection points account for the high percentage.

7. The Pacific Gas and Electric Company's wind farm is the largest in the world. It is located at Altamont Pass, California, USA, and covers 54 miles. Since 1981, its 7,300 turbines have produced over 6 billion kilo-watt hours of electricity, which is enough to power 800,000 homes for a year.

8. Switzerland has the highest per-capita solar energy usage in the world, with 1.82 watts per capita. In Germany 0.71 watts of solar power

is used per capita, while Japan is in third place in terms of solar power usage with 0.65 watts per capita.

9. Putting His Money Where His Mouth Is – On September 18, 1997 Ted Turner (USA), the founder of CNN and vice president of Time Warner, announced, he was giving the United Nations one billion dollars of his own money; this is the largest private donation ever to the UN. The money will be used to finance programs on behalf of refugees, children, the environment, and clearing landmines. Way to go, Ted!

10. Largest Federation of Environmental Groups – Friends of the Earth International is the world's largest federation of environmental groups ever organized. It unites one million activists throughout the world, with member organizations in 68 countries and affiliate groups.

Source: (1) Guinness World Records, 2002, Page 106

(2)Guinness World Records, 2003, (all the rest, Pages 64 & up)

Note to point 10 – Perhaps a large powerful international organization is needed to deal with large polluters, such as the oil and nuclear electric producing industries. I will be interested to see what this organization can do. There is no doubt, pollution has become a worldwide problem, and should be dealt with that way. The question is, will this group have any clout?

CHAPTER EIGHTEEN

THE SWELLING WORLD'S POPULATION

The Environmental Protection Agency was organized in 1970 as an independent U. S. government agency which controls solid waste, pesticides, noise and radiation pollution through research, development and training programs.

Their three most environmental concerns are:

1. Global warming – previously covered earlier in the book.

2. Poverty

3. Overpopulation.

For this chapter we want to concentrate mainly on points 2 & 3, poverty and over population, which we feel are closely related. We are inclined to agree with the EPA. The success of the human race is also their failure. The success: mankind has survived by the numbers i.e. population – over 6 billion and counting as of 8/16/02, estimation from the U. S. Census Bureau, growth rate 1.25 %, birth rate 21 births/1,000 population, death rate 9 deaths/1,000 population, sex ratio at birth 1.05 males/females, infant mortality rate 53 deaths/1,000 live births. The failure is: the world in its' current production of human needs, such as food, water, clothing and shelter/sanitation, is falling behind the increasing world population. The EPA estimates the earth will be at maximum saturation of population at 9 billion people, which at the current rate should be reached by the year 2012 estimated. That isn't too far away. Is the world prepared? To me it seems nobody cares whether we are or not; I don't believe there is even a worldwide game plan for that fateful day.

    Of course the EPA's woeful report wasn't the only prediction of overpopulation despair. (1) As far back as 1798, Thomas Robert Malthus, a young English clergyman, wrote a book entitled, Essay on the Principle of Population. It went through several editions and for a century influenced the thinking of people all over the world, including Charles Darwin. It is still a living influence today. Malthus' view depends directly on the law of diminishing returns, i.e. an increase in some inputs relative to other fixed inputs will cause total output to increase; but after a point the extra output resulting from the same additions of extra input is likely to become less and less. Malthus postulated, a universal tendency for population – unless checked by food supply – to grow at a geometric progression, becomes so large that there isn't space in the world for all the resulting off-springs to stand.

Source: (1)Economics – An Introductory Analysis, by Paul A. Samuelson, Professor of Economics, Massachusetts Institute of Technology, Fifth Edition, Publisher McGraw-Hill Book Company, Inc., New York, NY, 1961, Page 27.

So, it is clear two hundred years later, the cupboard isn't bare, the world is still spinning around, and still providing for its' population, or is it? Was Malthus wrong? He didn't take into consideration the Industrial Revolution or assembly line mass production. But, is there another Industrial Revolution, which will provide as the other did? I don't know of any. However, I do have some ideas on the matter of providing, which I'll discuss under the caption, "Water –

Provide For More People". Getting back to Malthus; perhaps he was right: the EPA seems to side with him on the population issue. A good way to examine the possibility is to review world poverty statistics, which I did. Here are some disturbing numbers that (2) the United Nations Human Development Report of 2002 reported concerning world poverty: the proportion of the world's extremely poor was at 23% of the world's population in 1999. 2.8 billion people live on less than $2.00 a day, with 1.2 billion of them surviving on the margins of subsistence with less than $1.00 a day. In 2000, 1.1 billion did not have access to safe water, and 2.4 billion did not have access to any form of improved sanitation services.

(2) Source: the U. N. Human Development Report 2002

Comments: That's a lot of poor people wouldn't you say? Perhaps the day has come when the earth can't provide for its' people. If that day is coming as the EPA suggests, then what can be done about it? Well there are two alternatives: eliminate people or provide for more people.

Let's look at eliminating people first: birth control (the pill) and death. What can I say about the pill, which hasn't already been said? Apparently, it hasn't been successful in curtailing population growth; the world has six billion people living in it, with no end in sight. The reasons I would surmise: people are ignorant about the procedure or just too lazy to go to the trouble, or don't have the money to buy the pills. Yes it does keep a sexually active woman from becoming pregnant; so you could say it does the job, but she must be devoted to it, day in and day out. You can draw your own conclusions from what I said. In my point of view it simply isn't working as a world population control.

That leaves the second alternative: death. Death has afflicted humanity by natural means i.e. natural disasters and pestilence, and manmade. The former two are outside man's control; the latter isn't. They include: homicide, suicide, auto accidents, wars, abortion, and genocide. Throughout the work, I've listed statistics on the first four, and I consider them upsetting. But for the limits of the book, I don't want to go further in detail and description of these evils. They are in themselves a subject for another book(s).

Instead I would like to limit my attention to abortion and genocide, since the two are still being practiced today, but in a somewhat subtle way. My partner and I feel this should be revealed, since it is in keeping with the nature and theme of the writing: to give you the truth, even though it may be upsetting.

Due to the length of both subjects, i.e. Abortion and Genocide, I'm going to give each a chapter of its' own.

CHAPTER NINETEEN

ABORTION

The U. S. Supreme Court in Roe vs. Wade, ruled 7-2 on 1/22/73 that states may not ban abortions during the first three months of pregnancy and may regulate, but not ban, abortions during second trimester. Simply put, they made abortions legal!

Source: The World Almanac 1999, Page 522

(1) What is Abortion? Abortion is the termination of pregnancy before the fetus is capable of independent life. When the expulsion from the womb occurs after the fetus becomes viable (capable of independent life), usually in the sixth month of pregnancy, it is called premature birth.

Abortion may be spontaneous or induced. Expelled fetuses weighing less than 14 ounces or of less than twenty weeks' gestation usually are considered abortion. It is estimated that 5 to 15 percent of all human pregnancies terminate spontaneously in abortion, with three out of four abortions occurring during the first three months of pregnancy. Some women apparently have a tendency to abort, and recurrent abortion decreases the probability of subsequent successful childbirth.

Therapeutic abortion is the deliberate removal of the fetus from the womb by means of instruments. The procedure is considered a surgical operation with all the precautions taken. Risks of hemorrhaging and infection are possible with each situation, and should be viewed with this in mind before a woman agrees to submit to it.

(1) Source: Funk & Wagnall's Encyclopedia, Pages 123 & 124

(2) Some facts about abortion:

48% of pregnancies among American women are unintended; half of these are terminated by abortion.

In 1997, 1.33 million abortions took place, down from an estimated 1.61 million in 1990. From 1973 through 1997, more than 35 million legal abortions occurred.

Each year, two out of every 100 women aged 15-44 have an abortion; 47% of them have had at least one previous abortion and 55% have had a previous birth.

An estimated 43% of women will have at least one abortion by the time they are 45 years old.

Each year, an estimated 46 million abortions occur worldwide. Of these, 20 million procedures are obtained illegally.

52% of U. S. women obtaining abortions are younger than 25; women aged 20-24 obtain 32% of all abortions, and teenagers obtain 20%

Black women are more than three times as likely as white women to have an abortion, and Hispanic women are twice as likely.

Catholic women are 29% more likely than Protestants to have an abortion, but are about as likely as all women nationally to do so.

Two thirds of all abortions are among never-married women.

On the average, women give at least three reasons for choosing abortion: three-fourths say that having a baby would interfere with work, school, or other responsibilities; about two-thirds say they can't afford a child; and half say they do not want to be a single parent or are having problems with their husband or partner.

About 13,000 women have abortions each

year following rape or incest.

(2) Source: Time Almanac, 2003, Page 557

CHAPTER TWENTY

ABORTION REALITY – GENOCIDE?

(1)Ban on Late-term Abortion Procedure Fails – The Senate on 9/28/98, sustained a veto by President Clinton of a prohibition on a late-term abortion procedure sometimes called partial-birth abortion. This was the second time the Senate failed to override a veto of the ban. The Senate vote was 64-36, 3 short of the two-thirds needed to override.

Source: The World Almanac 1999, Page 63

Author's Note: I've given the facts and statistics about abortion. By the numbers the procedure appears to be generally acceptable worldwide as well as in the United States. However the numbers also point out the disturbing fact that abortion is directed at the lower class, since most abortion clinics are located in poorer neighborhoods. Can I take from this that the practice is genocidal in nature at least on how it's practiced in the USA? Is America killing their poor population before they are born: the destruction of poor people, so the state won't have to deal with them; the undesirables, the people on welfare, the people who are destined to fill up the jails. After all, the aforementioned is an expensive burden on the state. Besides, the country doesn't need these people anymore: all of their work is now sent overseas; the Chinese and Indians do it, as well as other third world nations. Can I surmise that this is what abortion is all about: disposing of the undesirables before they become undesirables? Is it true that in America, when people are no longer needed, it is time to get rid of them? The Supreme Court in their Wade vs. Roe decision, appears to corroborate my suspicion in making this procedure legal.

Continuing on with the death theory of controlling population: can it be said that abortion worked? No! The population of the country is still rising and so is the poverty level. Since Roe vs. Wade, 43 million abortions have been performed.

I'll have more to say about genocide in the Chapter entitled GENOCIDE (The American Indian).

PHOTOS AT EIGHT WEEKS GESTATION

In a ground-breaking article in the April 30, 1965, issue of LIFE Magazine, Swedish photographer Lennart Nilsson revealed what it looks like inside the womb. (The photos were shown at eight weeks gestation, approximate size from head to toe: 1.5 inches and at 18 weeks gestation, approximate size: eight inches. The baby in the photos looks just like a baby, with a head, arms, legs and detailed facial features). His photos were a scientific and aesthetic marvel. "This is like the first look at the dark side of the moon," said one gynecologist. In the years since, Nilsson's images of fetal development have been used by both sides in the rancorous debate over abortion. But the man who has actually witnessed a sperm enter an egg, and who has photographed every stage of human gestation, declares that even he cannot determine when life truly begins. "Maybe," he says, with a smile, "it starts with a kiss."

Source: LIFE Magazine – 1965, Sixty Years, a 60th Anniversary Celebration 1936-1996, Page 101

CHAPTER TWENTY ONE

GENOCIDE (The American Indian)

The two most noted forms of attempted genocide were Hitler's Germany of the early 1940's (6 million estimated Jewish deaths) and The Mongol Empire's 35 million deaths of Chinese peasantry, 1311 – 1340. Both events are indeed appalling and worthy of more consideration, which I won't be able to give here. It is enough that they took place, putting a terrible blight on human history. My point in mentioning them is: genocide isn't anything new. It is a human characteristic that doesn't seem to go away; practiced by many empires and cultures all over the world. Unfortunately it hasn't gone away from America either. The American Indian will attest to this fact. He will of course, because he was victim to it, but unfortunately no one else will. America just wouldn't do such a thing as genocide; it is ugly and anti-American. It simply can't be; America is always the good guy. They wouldn't do anything as despicable as killing off an entire race of people. But they did! They should be held accountable for it, at least in writing if nowhere else. I would prefer some sort of restitution be given to these people (what's left of them) for the great injustices they and their ancestors suffered at the hands of the United States. But I don't see that happening in the near future, or maybe never.

And so I can only give the truth and let it dangle in space like some ugly thing, which it is. The following are some of the facts about our red brother, whom we mistakenly dub, "The American Indian."   (Columbus, when arriving in America, wasn't aware of where he was. Consulting his charts, the explorer was convinced that he was in the Indies and called the natives "Indians").

The American Indian was the first to arrive on the continent. He was kind to the environment, living off the land: didn't overuse it, allowing it to naturally fertilize itself by moving his planting fields; and ate all he killed; not slaying animals simply for the sport of it. When the white man arrived, he found a pristine part of this earth, which could support vast millions of people, if used properly. The landscape from ocean to ocean was magnificently beautiful; poets had a field day describing it: "...my country tis of thee, sweet land of liberty, of thee I sing, from sea to shining sea, and so forth." Unfortunately for the Indian, the arrival of the Europeans in the Americas meant this 'land of liberty' would be for the white man only; not to be shared with the American Indian. He had no future with the new white culture, which would dominate the land. He simply didn't fit into it. In fact the white man had something else in store for his red brother: annihilation.

In doing my research, I came across an old college history text book of mine, which has a passage in it describing how the white man treated the American Indian. It made me sick, when I read it. (1)The book was entitled, "The American Nation – A History of the United States," by John A. Garraty. The following is a summary from a section of the text entitled, "Indian Wars," which illustrates the white man's lack of integrity in his dealing with the Indian.

The American government and people rarely honored treaties made with Indians. For example, when the Kansas-Nebraska Bill became law, tribes like the Kansas, Omaha, Pawnee, and Yanton Sioux began to feel the pressure for further concessions of territory. Once the Kansas and Nebraska territories had been cleared (about 1860); the Indians had lost all but 1.5 million of their 19-odd million acres. The Colorado gold rush in 1859 sent thousands of greedy prospectors across the plains to drive the Cheyenne and Arapaho from lands guaranteed them in 1851. In the Sioux country other trouble developed. In 1862, after federal troops had been pulled out of the west for service against the Confederacy, many of the plains Indians rose up against the whites. For five years intermittent but bloody clashes kept the whole area in a state of alarm.

Guerrilla warfare developed, with all its horrors and treachery. The Chivington Massacre of 1864 occurred, where a party of Colorado militia fell upon an unsuspecting Cheyenne community at Sand Creek and murdered an estimated 450 Indians. Colonel J. M. Chivington, a minister in private life, told his men, "Kill and scalp all, big and little. Nits make lice." A white observer of the scene described what happened, "The Indians were scalped, their brains knocked out; the men used their knives, ripped open women, clubbed little children, knocked them in the head with their guns, beat their brains out, mutilated their bodies, etc." Afterwards General Nelson A. Miles called it the "the foulest and most unjustifiable crime in American history."

The Indians retaliated by wiping out dozens of isolated white families, ambushed small parties, and fought many successful skirmishes against troops and militia. Their most notable triumph was in December, 1866, when the Oglala Sioux, under their great chief Red Cloud, completely wiped out a party of 83 soldiers under Captain W.J. Fetterman. The Indians under Red Cloud fought ruthlessly, but only when pressured by the construction of the Bozeman Trail, a road through the heart of the Sioux hunting grounds in southern Montana.

As a result the government in 1867 evolved a new strategy. Their 'concentration' policy hadn't gone far enough. Under their new directive the plains Indians would be confined to two small reservation, one in the Black Hills of Dakota Territory, the other in Oklahoma. The idea was to force them to abandon their wild habits and become farmers. At two conclaves held in1867 and 1868, one at Medicine Lodge Creek and the other at Fort Laramie, the principal chiefs agreed to the government's impositions (or else).

Regardless, a good many of the tribesmen wouldn't abide by these agreements. They would rather die than give up their way of life. Once again, they raged again across the plains, like a prairie fire and fought the white man. General Philip Sheridan, Grant's great cavalry commander, made the following comments about the situation: "We took away their country," he said, "and their mode of living, their habits of life, introduced disease and decay among them, and it was for this and against this that they made war. Could anyone expect less?"

(1) Source: The American Nation–A History of the United States By John A. Garraty, Harper & Row, Publishers, Inc. – American Heritage Publishing Co., Inc., New York – 1966 – Page 477

General Sheridan's comments just about sum up how the white man treated the American Indian across the nation. He kept pushing and pushing the red man, until they couldn't go any further, and then locked them into various reservations on worthless pieces of land to live in poverty and degradation. They are remembered only as nicknames for major league sports teams: Kansas City Chiefs, Washington Redskins, Cleveland Indians, Atlanta Braves, Chicago Black Hawks to name a few. Their real legacy, which is lost in American history, was their respect for the environment. In reality the Indian had a better understanding of life than the white man with all his wonderful inventions, which are now polluting the earth, we depend on for survival. Did anyone listen? No! If there is something redeeming in this tragedy, you could say the American Indian had the last laugh, but none of this is funny. And what happened to the red man isn't funny. Let me give you some numbers, which will give you an idea on how he fared under the white man's system.

The following is a list of the ten largest Indian reservations and the population of each:
Navajo (Ariz., N. M., Utah)   143,405
Pine Ridge (Neb., S. D.) 11,182
Fort Apache (Ariz.)   9,825
Gila River (Ariz.) 9,116

Papago (Ariz.) 8,480

Rosebud (S. D.) 8,043

San Carlos (Ariz.)   7,110

Zuni Pueblo (Ariz., N. M.)   7,073

Hopi (Ariz.)   7,061

Blackfeet (Mont.) 7,025

Note: The 218,320 American Indians living on these reservations account for about half of all Indians living on reservations and trust lands in the United States.

Source: 1990 Census Bureau

Compare this to the total population of the United States, which comes to 281,421,906 (Source: U. S. Census bureau, Census 2000). What does that tell you?   Genocide?

This is Webster's New Collegiate Dictionary's meaning of the word: "the deliberate and systematic destruction of a racial, political, or cultural group." Sounds like it fits to me!

Personal Opinion: This sums up for me the option of "Killing" as a method of controlling world population. Ask yourself, "has killing done anything to solve this problem?" No, the world population is still growing at an alarming rate. All killing does is to show mankind for what he really is: an uncivilized savage!

Maybe, 'providing for the growing population,' is the better way to approach the overcrowding issue. As I mentioned before "Water", might be the answer. That's coming up next with a list of America's most endangered rivers..

CHAPTER TWENTY TWO
AMERICA'S ELEVEN MOST ENDANGERED RIVERS

1. Missouri River – Dam operations.
2. Big Sun flower River – flood control projects.
3. Klamath River – Water withdrawal and pollution.
4. Kansas River – Pollution; removal of Clean Water Act protection.
5. White River – Navigation and irrigation projects.
6. Powder River – Coal bed methane extraction.
7. Altamaha River – Reservoir and power plant construction.
8. Allagash Wilderness Waterway – Removal from the Wild and Scenic Rivers System; loss of wilderness values.
9. Canning River – Oil and gas exploration and development.
10. Guadalupe River – Water diversion.
11. Apalachicola River – Navigation and water withdrawals.

Source: Time Almanac 2003, Page 597

CHAPTER TWENTY THREE

WATER (INTRODUCTION – PROVIDE FOR MORE PEOPLE)

"...Wherever the stream flows, there will be all kinds of animals and fish. The stream will make the water of the Dead Sea fresh, and wherever it flows, it will bring life. ...On each bank of the stream all kinds of trees will grow to provide food."

Source: Ezekiel 47 9–12 Old Testament of the Holy Bible.

The above isn't a fairytale; it can be done scientifically. All that has to be done is to stop polluting the rivers, lakes and streams. This can be accomplished with responsible leadership and people who are educated on the environmental problems, and are willing to make a concerted effort to do something about it. Am I asking for too much? Maybe I am! But regardless, allow me to make some suggestions on how it could be done.

1. Irrigation is a proven possibility. Simply put, it is the application of water to farmlands by

artificial means. It is practiced all over the world, where the rainfall is insufficient to provide water for plant growth. In certain areas of the southwestern U. S., there isn't enough rain during the year to permit crop raising; in other areas the rainfall is insufficient during the growing season.
Additionally growers of high-priced crops use irrigation as insurance against period of drought.

    Two techniques are available in the distribution of irrigation. The water may be poured over the entire cultivated area, allowing it to stand until the ground is thoroughly soaked and then drained away. Or the water may be led through the fields via small parallel channels or furrows from which it can soak into the soil. The former system is adaptable to the use of flood waters, which usually occur once a year. However, the system using channels is far more flexible and far more common.

    Drainage is of the first importance in any irrigation system. Most natural surface fresh water contains a certain amount of dissolved salts, which become harmful to plant life, when they are concentrated. Provisions must be made for the drainage of irrigated fields, or the soil will soon become saturated with salts and become worthless for agriculture. The exception is in the case of loose or sandy soil, which provides natural drainage. Good irrigation practice demands the construction of suitable drain channels, usually set parallel to the irrigation channels and between them.

2. Amount of Water – The rule of thumb is 300 to 500 pounds of water is needed annually to produce one pound of dry crop. In addition to this minimal requirement, allowance must be made for evaporation, since the percolation of water in the ground may go below the growing zone of the crops. The entire amount of water that is needed by the plants and to offset the losses inherent in the nature of the soil, the type of terrain, and the irrigation system itself is referenced as the 'duty of water'. This may vary within wide limits, but the figure of one hundred acres per second-foot is often taken as an average in the irrigated areas of western U. S. Values of farm land in irrigation districts is decided by the duty of water and it's supply. The principle irrigated farm products are forage crops, potatoes, cotton, rice, sugar cane, fruits, berries, and grains. Irrigated

farms produce almost twice the average farm yield. With other methods of soil management, irrigation may yield three times the average farm production.

Source: Funk & Wagnall's Encyclopedia, Standard Reference Works Publishing Company, Inc., New York, NY, Page 5036

ISRAEL

Israel's water supply is severely limited. Farming water in many cases is drawn from rivers, springs, underground pools or other channels to areas under cultivation. The Negeb region is a sparsely inhabited desert in southern Israel; the rainfall there averages less than ten inches annually. The area is irrigated by water piped about 63 miles south from the Yarkon River, which flows into the Mediterranean near Tel Aviv. Yarkon-Negeb water pipelines were erected there: the one measures 66 inches in diameter; the other has a diameter of 70 inches. The estimated capacity of the two lines is about 260,000,000 cubic yards of water per year. Other water for irrigation is obtained from wells in the foothills of Galilee, from the Kishon River, and from Lake Huleh, in the highlands. Israel cannot make extensive use of the Jordan River for irrigation, because of disputes with neighboring countries.

Regardless, the Israel state has transformed thousands of arid acres into fertile fields by piping water many miles from its source for irrigation. The country, through this initiative, has developed their agriculture to the point of abundance; their most important crop is citrus fruits. Other principal crops are olives, barley, wheat, tomatoes, millet, potatoes, figs, corn, and durra. Their best farming regions are Esdraelon, or the Plain of Jezreel, in the northeast and Sharon, the plain along the Mediterranean coast.

Jewish farming in Palestine occurred almost entirely after the beginning of the twentieth century; before that there were only a few scattered Jewish settlements. Jewish farming settlements were created by various Zionist organization on parcels of land purchased from the British government of Palestine or from Arab landowners. These settlements are organized into three principal types, and were colonized by Jewish immigrants of limited farming experience. The most recognized type of farm colony is the communal or cooperative settlement. Here the workers live in communal dwellings, and share the work and profits equally. The second type is called the worker' smallholders' settlement, where individual farms are worked separately. The produce from them is pooled and marketed by the settlement unit. Both types of settlements are established on land owned by one of the various colonial organizations, and then is leased to the settlers. The principal one is the Jewish National fund. The third type is the smallholders' settlements, where individual farms are worked as private enterprises. In addition in recent years, several other types of agricultural settlements have evolved, which intermediate between the collective and the free-enterprise communities.

    Source: Funk & Wagnall's Encyclopedia, Standard Works Publishing Company, New York, NY, Page 5054

    Author's Note: If a small state such as Israel can be so successful in providing food via irrigation, just imagine what could be accomplished doing the same thing with the Sahara desert running pipelines of water from the mighty Nile River. Just imagine, if you will, and see the possibilities!

To do so it may help, if I give you some statistics about the two in the next chapter. The measurement and scope of them may astound you. It did me.

CHAPTER TWENTY FOUR
SAHARA DESERT & THE NILE RIVER

The Sahara desert of northern Africa extends from the Atlantic Ocean eastward past the Red Sea to Iraq. The desert varies in width from 1400 to 800 miles and is about 3200 miles in length from east to west at latitude 20 degrees north. Its' total area is more than 3,500,000 square miles: approximately 80,000 miles consist of partially fertile oases. The boundaries of the desert are not clearly defined. They are generally accepted to be the Atlantic Ocean on the west, the Atlas Mountains and the Mediterranean Sea on the north, the Red Sea and Egypt on the east, and the Sudan and the valley of the Niger River on the south.

Three divisions of the Sahara are geographically distinct: the Western Sahara, known as the Sahara proper, the central Tuareg Plateau, and the Libyan Desert in the east. The western part of the desert is an area of rock–strewn plains and sandy deserts of varying elevation. It is almost entirely without rainfall or surface water. However it possesses a number of underground rivers, which flow from the Atlas mountains and the mountains of the Tuareg Plateau. In some instances the waters from these rivers find their way to the surface creating naturally irrigated oases, where plants grow freely. The soil from this region of the Sahara is naturally fertile; where irrigation is possible, excellent crops can be produced. The Sahara central plateau region runs for approximately 1000 miles in a northwest to southeast direction. The plateau varies in height from 1900 to 2500 feet and peaks in the several mountain ranges. Although the rainfall is scarce, a few of these peaks are snow-capped during part of the year. The Libyan Desert is the driest part of the Sahara. Moisture is almost totally absent and few oases exist. The land is mostly sandy wastes, with large dunes of 400 or more feet in height. The valley of the Nile and the mountainous area of the Nubian Desert to the east are geographically a part of the Sahara. The irrigation afforded by the Nile transforms the desert into fertile agricultural land throughout much of Egypt.

Source: Funk & Wagnall's Encyclopedia, Standard Reference Works Publishing Company,

New York, NY, Page 7750

THE NILE RIVER

If the Sahara desert is a formidable tract of land, the Nile River is equally formidable in size producing an abundance of water, which as I stated before might be capable of irrigating the world's largest desert, using the Israelites' system as a prototype.

The following is a breakdown of the river's dimensions. The Nile is the longest river of Africa. It draws its water from the Victoria Nyanza and Albert Nyanza lakes. Its source is one of the upper branches of the Kagera River, which originates in the mountains of Tanganyika. On leaving the Victoria Nyanza Lake it pours over the Ripon Falls, 170 yards wide but only 12 feet high, and then for 300 miles rushes between high rocky walls, over rapids and cataracts, at first northwestward, then westward, until it joins the Albert Nyanza Lake. The section between the two lakes is called the Victorian Nile or Somerset River. At its southwestern extremity the Albert N. it is joined by the Semliki River, which drains the surplus water of the Albert Edward

Nyanza. This lake drains the slopes of the snowy Ruwenzori Mountain and other adjacent mountains. The combined Somerset and Semliki rivers leave the northern extremity of the Albert Nyanza and become the Bahr el Jebel River. Various tributaries flow through the Bahr el Ghazal district.

The White Nile River, or Bahr el Abyad at Khartoum is joined by the Blue Nile River. The water of the respective streams are the color indicated by their names. The region adjoining the White Nile River above its link up with the Sobat River was for a time known as the Equatorial Provinces. The Blue Nile River, which is 950 miles long, gathers its volume principally from Lake Tsana, on the Abyssinian plateau, this region is known as the Abbai. From Khartoum the Nile flows northeastward and 200 miles below that city is joined by the Black Nile River. The black sediment brought down by this river settles in the Nile Delta, which makes it extraordinarily fertile. During the river's course from the confluence of the Black Nile River through the Nubian Desert, the great river makes two deep bends. From here (below Khartoum) navigation is rendered dangerous by cataracts; the sixth occurring north of

Khartoum, the first near Black Nile River. The Nile then enters the Mediterranean by a delta, which separates into two main channels, known as the Rosetta and the Damietta. Its total length is about 4000 miles, or 3470 miles from its outlet in Victoria Nyanza. Its basin has an area of over 1,000,000 square miles.

Source: Funk & Wagnall's Encyclopedia, Standard Reference Works Publishing Company, New York, NY, Page 6548

Author's Note: Now if you are interested in sailing down the Nile River, you'll know where you're going. All of this aside, my point here is to give you an idea of what this great river is all about, and how much of northern Africa it covers.

Consider this, if oil can be piped from Alaska, through Canada to the United States, can't water be piped from the Nile River throughout the Sahara Desert. Along with this, wells are dug tapping the underground rivers of the desert to provide the necessary water to irrigate the land. I believe it can, and by so doing feed the entire African continent with leftovers for export. In turn the land of the Sahara can be opened to support more of the earth's peoples. I haven't mentioned the jobs created by the project, and future jobs to run the farms and the commerce, which would follow. Africa would become economically sound, and take its' place among the elite nations of the world. It could happen. All that is required is a desire to do it, along with an ending of political maneuvering by power hungry small minded leaders, working against one another. To achieve great deeds, great men must come forth capable of moving effectively in a direction beneficial to all mankind.

CHAPTER TWENTY FIVE
WATER POLLUTION BY PEOPLE

People living too close to the water: people have a tendency to live close to water i.e. rivers, lakes, bays, oceans, etc. What comes of this is a greater possibility of polluting the water from their housing developments. In any development it will cover the soil with concrete and asphalt thereby creating a runoff of wastes such as fertilizer, dog excrement, etc. into the water. Without soil to absorb this, it has only one way to go (to the water).

This is more serious than it sounds. There have been reported instances where non – biodegradable materials, such as a plastic bag was found in a whale's stomach blocking the animal's intake of food, causing the beast to die of starvation. Of course, the fish life in general in the water is compromised, along with the use of the water for human bathing.

End of: WATER

CHAPTER TWENTY SIX   FOOD AND DIET

Note: Understanding food and what to eat and not to eat can change your life. I believe for the better. The facts I'm going to present to you in the following 25 pages will illustrate this and hopefully adjust your way of thinking, especially about the common popular foods. Literally you are poisoning yourself and your children eating these foods, with the Food and Drug Administration standing by allowing it to happen.

To support my findings, I've tapped the minds of many qualified research scientists. To them I give my full appreciation, and acknowledge that without this information the work would be lacking.

CHAPTER TWENTY SEVEN

BAD DIET LINKED TO ALCOHOLISM

Addiction, especially alcoholism, has plagued mankind for a long time, dating back probably from the beginning. Although we all recognize this and the perils of addiction, many fall into its' trap. We all think of curing it by simply stopping (cold turkey). But is their another way? Perhaps a way right under our noses. What about nutritional measures in treating compulsive drinking?

(1)Dr. Roger J. Williams, University of Texas biochemist, and one of our country's leading nutritionists, has noted in his research that although alcoholics tend to have a poor diet (often through neglect), the reverse situation also applies: bad diet may create alcoholism.

Dr. Williams came up with the idea that alcoholism is essentially a disease of a person's appetite and can be consistently prevented by the application of nutritional knowledge. He concluded in his classic work, Alcoholism: The Nutritional Approach (University of Texas Press, 1959) that all people have a potential in them of becoming addicted. He cited body chemistry and the individual's like for alcohol as factors of contracting the disease, and warns that the person should be on his guard, always, at all times.

Dr. Williams has been studying the problem of alcoholism and nutritional disorders for many years. He came up with the following: alcohol contains large amounts of carbohydrates and no vitamins or minerals – thus making it very difficult for the body to utilize the carbohydrates found in the alcohol. Consequently, if an individual develops a drinking habit, in time the alcohol consumed uses up the vitamins and minerals stored in the liver for carbohydrate processing.

Williams In his studies further concluded by using rats extensively that the tendency of experimental animals to drink alcohol is influenced most remarkably by the composition of the food they get. He explained that the deficiency of the B complex vitamins caused by drinking alcohol, created a greater need for glucose. Also the lack of the B complex vitamins induced malfunctioning of the liver. Both malfunctions contributed to liver injury and interfered with its repair. This liver injury, Dr. Williams concluded, may also contribute to the craving for alcohol.

By drinking, an individual has given himself a deficiency in a number of important B vitamins. This makes it impossible to convert glucose into its storage form (which is glycogen). Being unable to convert surplus glucose and thus remove it from the bloodstream until needed, the liver is failing to function properly and the sugar piles up in the blood. The body then responds to this excess glucose by releasing additional insulin into the bloodstream. Then as the body needs more glucose, the liver is unable to supply it. Hypoglycemia – low blood sugar – sets in and becomes chronic in the individual. Lacking glucose, energy is low and the brain does not function properly. The alcoholic feels he needs the quick energy he gets from alcohol, as the only thing that will set him right. He may actually believe, as many alcoholics do, that he will die if he doesn't get a drink soon.

From his rat research, Dr. Williams discovered that they craved alcohol instead of craving good food. This proved the theory that a B complex deficiency manifests itself as a disorder in the body's appetite mechanism. Furthermore, he concluded that body wisdom becomes body foolishness in the rat, because of the irregular appetite mechanism,

  (1) Source: Alcoholism: Nutritional Approach (University of Texas) 1959

  (2) A similar study was conducted at Loma Linda University in California, 1967 by Dr. U. D. Register. Instead of artificially manipulating the diet, Dr. Register experimented with feeding rats a diet, which is common among people in the United States. The idea of the experiment was to see if an individual's normal diet of food and drink affected his rate of alcohol consumption.

  Dr. Register and his colleagues believed the popular, but poor quality American diet consisted of hot dogs, spaghetti, meat balls, sweet rolls and soft drinks. Using this as their guideline, they fed this to their test rats. Also the rats were given the choice of drinking either ten percent alcohol or water. The popular diet was compared to the standard good nutritional food prepared in the laboratory consisting of corn-soy-alfalfa diet – and a control "milk-vegetable" diet.

Their findings concluded that the rats being fed the popular diet consumed 12.8 ml. of alcohol per 100 grams body weight per week, as compared to the rats on the milk-vegetable control diet, who were consuming 2.3 ml. of alcohol. Adding coffee to the popular diet, doubled the alcohol consumption. When the diet was augmented with both coffee and spices (such as cloves and mustard) the alcohol consumption of the rats rose to 46 ml. per week.

The rats, after being on the U. S. diet for ten weeks, were switched to the control diet. During the first week, the drinking level dropped to 3.8 ml. and within three weeks, most of the rats had completely stopped drinking. Indicative of how strong the relationship to diet is, when the "popular diet" was resumed, the rats returned to the previous level of alcohol consumption.

Dr. Register in concluding his report, found the deficiencies resulting from the lack of the vitamin B complex family may have accounted for the basic dependency that the drinker has on alcohol. Without these vitamins, humans become quite confused, irritable, unstable, and are unable to adjust to many stressful situations. He also concluded, the combination of the popular poor quality U. S. diet, combined with coffee and spices, could very well lead a person to alcohol as an anesthetic or a poor solution to the increasing problems facing him in his stressful, complex life.

The accepted belief that an alcoholic is a person, who simply won't "pull himself together" and stop drinking is now in question. Perhaps, as Dr. Register's research suggests, the alcoholic is unable to stop drinking, because he can't – just as he can't stop using air and water. A definite physiological craving has been created for alcohol by a deficiency of the Vitamin B Complex.

(2) Source: Encyclopedia of Common Diseases by The Staff of Prevention Magazine – Rodale Press, Inc. 1976

CHAPTER TWENTY EIGHT
FOOD AND DRUG ADMINISTRATION

The Food and Drug Administration is a division of the U. S. Department of Health, Education and Welfare. In 1953 it was transferred to and became a part of the Federal Security Agency. The Food and Drug Administration or FDA, is charged with the enforcement of the Federal statues (i.e. Pure Food and Drug Acts) regulating the purity, standards of potency, and labeling of foods, drugs, and cosmetics, in interstate commerce.

The FDA does intensive studies concerning the composition of food products, trade practices, and consumer attitudes. From these studies they formulate definitions and standards of purity, and evaluate the safety for public use of new drugs, issuing or withholding permits for their distribution. The FDA makes special certifications in advance for insulin, penicillin, streptomycin, and coal-tar colors used in foods, drugs, and cosmetics. They do this because certain factors in their manufacture may render inadequate the

FDA's customary enforcement procedures, which are described below. Previous to shipping, samples of each batch of these drugs and colors are submitted to the FDA for approval of their purity and potency.

The FDA has special laboratories in Washington, D. C., where they make investigations to perfect methods of detecting and proving *adulteration of food, drugs, and cosmetics. These studies are two types: fundamental research to acquire data for formulating general policies; and to determine specialized techniques and apparatus, which aren't available in the station laboratories. The fundamental research evaluates the following:

1. The safety and efficacy of medicines;

2. The toxicity of the ingredients used in the manufacture of the food, drugs, and cosmetics;

3. The potency of the food, drugs, and cosmetics;

4. The potency of the drugs and vitamins;

5. The investigations of methods of processing, packaging, preserves, and storing products.

The specialized techniques include vitamin

and antibiotic assays, pharmacological bioassays, bacteriological analyses, and intricate microscopic determinations.

Source: Funk and Wagnall's Encyclopedia, Standard Reference Works Publishing Company, New York, NY, P. 3606

*Adulteration is the act of making any commodity impure by mixture of other or baser ingredients. This added mixture may corrupt the nature of the original base to the extent of destroying its' identity, or may lower the value of the efficacy of the finished product. The object of adulteration is to increase profit for the manufacturer, who sells inferior food, drugs, and other products to an unwitting public. Such products and their misbranded labeling of standard quality, may constitute a grave menace to public health.

In 1906 the Federal government enacted the National Pure Food and Drug Act. Within a few years of the passage of the Federal statute, shipping, inspection, and proper branding and labeling laws intended to insure maximum Purity, were put into place. State laws regulating the production and sale of food and drug products within their borders widened the scope of the Federal law, which could attack the problem only through the power of Congress to regulate interstate commerce.

The Federal Food and Drug and Cosmetic Act of June, 1938, went further to decrease incentive to adulterate and defraud. Penalties were increased, loopholes in the previous law were eliminated. Cosmetics were included among items under Federal control, and the regulations, regarding interstate traffic and labeling drugs, were made more stringent. In 1939 enforcement of the food and drug laws were placed under the Food and Drug Administration. Previous to that they were under the Department of Agriculture.

Source: Funk and Wagnall's Encyclopedia, Standard Reference Works Publishing Company, New York, NY 1963, page 194

Author's Note: Now that we know what the FDA does or what they should do, let's see if they are doing their job.

CHAPTER TWENTY NINE

(1) FOOD ADDITIVES – CANCER CAUSING CHEMICALS

There are over 3,000 chemical additives that are purposefully added to the American food supply. Some of these are known to cause cancer. This is in spite of our country's laws forbidding such food treatment. Clearly, the Delaney Clause of the 1958 Food Additives Amendment deals with the subject: "chemicals, which have been shown to cause Cancer in humans or animals are not permitted to be added to commercially sold food." It sounds understandable to me, but not to the FDA. They decided not to enforce the law, and allow small amounts of the Cancer causing substances to be used in foods. I guess they must have been pressured by somebody important to permit this.

There is more. Not only do additives cause carcinogenic problems, some can bring on allergic reactions as well. There is even more: people in certain health group categories, such as infants, pregnant women, and those with kidney or heart problems are at risk from them.

It doesn't end there. Generally, foods contain more than one additive. Safety testing has been done on individual chemicals, but not on various additives in combination. To date there has been little research in this area. But, one has to be concerned with this, and consider it a red flag, since there are so many combinations of additives used in today's food products. So following along with this thought, what can the average person do in planning which foods are safe to eat? Somewhat of a help in this quest is: packaged food labels must list ingredients. However, finding and reading the ingredients may be problematic. The complication of reading and identifying the various ingredients is a task the average person isn't likely to do, and if he does, may have trouble doing so. So, I'm going to give the normal Joe or Jane some quick fix ideas on how to read ingredient labels. If the ingredient list is lengthy, there are most likely many additives in the product. It may be wise to avoid these foods due to the unknown risks of consuming combined chemicals. Label ingredients are listed by weight with the heaviest first and the lightest last. Questionable additives are usually towards the top of the list, and can be considered the most harmful.

The following codes indicate the safety of the additives in the table below. Many additives have more than one code used to describe their safety.

* Recognized as safe by the FDA.
# FDA approved colorant.
S There is no known toxicity.
A The additive may cause allergic reactions.
C Caution. The additive may be unsafe or poorly tested.
C1 Caution for certain groups of the population, such as pregnant women, infants, person with high blood pressure, kidney problems.
X Unsafe or poorly tested.

Some Common Food additives:
...X Acesulfame K – "Sunette"; may lead to low blood sugar; also causes cancer, and elevated cholesterol in lab animals.
...X Acesulfame-potassium – also known as Acesulfame-K.

...C Animal or vegetable shortening – contributes to heart disease, hardening of the arteries, and elevated cholesterol levels.

...X A Artificial color FD & C, US certified food color – add to hyperactivity in children; may affect learning and visual disorders; lead to nerve damage; and may be carcinogenic.

...X A Artificial flavoring – reproductive disorders, developmental problems; not adequately tested.

...X Artificial sweeteners – associated with various with health problems; refer to a specific sweetener.

...X Aspartame – brain damage possibilities; may cause central nervous system disorders, menstrual problems; may affect brain development in unborn fetus.

...X A BHA – may cause liver and kidney damage, behavioral problems, infertility, immune system weakening, birth defects, cancer; should be avoided by infants, young children, pregnant woman and people sensitive to aspirin.

...*X A BHT – refer BHA; banned in England.

...X A Brominated Vegetable Oil – connected to major organ system damage, birth defects, growth difficulties; considered unsafe by the FDA; however can still lawfully be used pending further action taken by the FDA.

...* X Caffeine – addictive drug; may contribute to fertility problems, birth defects, heart disease, depression, nervousness; may affect behavioral and sleep patterns.

... F D & C Colors – ruled safe by the FDA for use in food, drugs and cosmetics; many of these colors are from coal tar derivatives, and must be certified by the FDA not to contain more than 10ppm of lead and arsenic. This certification does not delineate harmful effects the colors may have on the body; many coal tar colors are potential carcinogens, and may contain carcinogenic products, and cause allergic reactions.

...X Free Glutamates – may cause brain damage, especially in children. Found in autolyzed acid, hydrolyzed protein, hydrolyzed soy protein, plant protein extract, protease, protease enzymes, and a wide variety of additives of this nature, and anything that is enzyme modified, fermented, protein fortified or ultra –pasteurized and foods that advertise NO MSG; see MSG.

...X A Hydrogenated Vegetable Oil – heart disease, breast and colon cancer, atherosclerosis, cholesterol elevation.

...X A Hydrolyzed Vegetable Protein – brain and nervous system damage in infants; high salt content. Contains free glutamates.

...X A MSG – headaches, itching, nausea, brain, nervous system, reproductive disorders, high blood pressure are related to its' use. Pregnant lactating mothers, infants, small children should avoid MSG; allergic reactions are common. It may be hidden in infant formula, low fat milk, candy, chewing gum, drinks, over-the-counter medications, especially children's binders and fillers for nutritional supplements, prescription and non-prescription drugs, IV fluids given in hospitals, chicken pox vaccine. It is sometimes sprayed on growing fruits and vegetables as a growth enhancer; and is proposed for use on organic crops.

...X A Natural Flavors – can be processed and combined with other food additives not required to be listed on the label; may contain free glutamates; see MSG.

...X Nitrates – from a combination with certain chemicals in the stomach, can change to Nitrites, a cancer-causing agent. Regarded as dangerous by

FDA, but not banned because it prevents botulism.
...X Nitrites - may cause headaches, nausea, vomiting, dizziness, cancer; see nitrates.

(1)Source: Food Additives - Protect Your Family From Cancer-Causing Chemicals! Copyright 2003-2007, HealthyEatingAdvisor.com All rights reserved. The Healthy Eating Advisor, Escondido, CA. Pages 1 to 4

More Information concerning Food Additives:

(2)Janet Starr Hull, creator of the Aspartame Detox Program has the following to report to the network entitled, Food Additives to Avoid:

As stated earlier, there are over 14,000 man-made chemicals added to our American food supply. Food additives aren't natural nutrition for humans, especially to the degree given in the common everyday diet. Children are most at risk from food additives, because they are exposed to them at a time when they are most vulnerable to the adverse effects of these chemicals.

Starr-Hull goes on to suggest that it is important for everyone to be aware of the types of chemicals and food additives they are consuming. She selected a few chemical food additives listed below as examples, but recommend the average person use the internet resources to research the myriad of chemicals and food additives inundating our modern food supply.

The following is her list:

...Olestra – Olestra is fake fat approved by the FDA, which is both dangerous and unnecessary. The additive was approved over the objection of dozens of leading scientists. Although fat-free, there is a fatal side-effect: it attaches to valuable nutrients and flushes them out of the body. Some of these nutrients – called carotenoids – protect against such diseases as lung cancer, prostate cancer, heart disease, and macular degeneration. The Harvard School of Public Health concurs with this and suggests that the long-term consumption of Olestra snack foods might result in several thousand unnecessary deaths each year from the cancers mentioned above. In addition, blindness in the elderly is possible due to macular degeneration. Besides contributing to disease, Olestra causes diarrhea and other serious gastrointestinal problems, even at low doses.

Olestra was certified by the FDA despite the fact that there are safe low fat snacks already on the market. No significant evidence is available linking Olestra to reduction of obesity.

Despite being approved as safe by the FDA, all snacks containing Olestra must carry a warning label (similar to the one found on cigarettes) citing possible health risks.

...Potassium Bromate is an additive, which has been used to increase the volume of bread and to produce bread with a fine crumb (the non-crust part of bread) structure. Most of this chemical rapidly breaks down to form benign bromide. However, bromate in itself has been known to cause cancer in animals. Although the amounts of the chemical that may remain in bread pose a small risk to consumers, bromate has been banned worldwide with the exception of Japan and the United States. It is rarely used in California because a cancer warning is required on the label.

Sulfites are a class of chemicals that can keep cut fruits and vegetables looking fresh and also prevent discoloration in apricots, raisins, and other dried fruits. In addition they control "black spot" on fresh caught shrimp; and prevent discoloration, bacterial growth, and fermentation in wine. Up until the early 1980's, they were considered safe for all foods. However, the CSPI (Center for Science in the Public Interest) has uncovered six scientific studies proving sulfites in some cases can provoke severe allergic reactions. CSPI together with the FDA identified at least a dozen fatalities linked to this chemical. All of the deaths occurred among asthmatics. As a result congress (1985) finally forced the FDA to ban sulfites from most fruits and vegetables. The asthmatic should be aware that his attacks might be related to sulfites, and regulate his diet accordingly. The ban does not cover fresh-cut potatoes, dried fruits, and wine.

(2)Source: Dr. Janet Starr Hull's newsletter, Aspartame dangers Revealed / Disclaimer / Link to us / Contact/ Site Map / Search. Copyright 2002, SweetPoison.com. All rights reserved. Pages I to 3.

End of Food Additives to Avoid

Author's Note: To answer my own question about the FDA doing their job properly: obviously not!

Now if you think the above is bad, wait until you read more into the chapter about the Metabolic

Syndrome, which I received through the internet, provided by the Mayo Clinic. It contains a great deal of important information, which if adhered to, could improve a good many diets and gain longer life expectancy and quality of life for those who follow the diet. And in addition , it will give notice that the FDA could do a better job of protecting the public against profit hungry food companies, who will do anything to make a buck, not caring who they hurt in the process.

Before getting too deep into the Metabolic Syndrome, it would be helpful to understand the meaning of certain medical terms: Pancreas, Insulin and Glucagon.

Pancreas: The pancreas is a gland, lying transversely in the posterior wall of the abdomen. It secretes a limpid, colorless fluid, which acts to digest protein, fats and carbohydrates. Also, internal secretions are produced by the Pancreas, namely insulin and glucagon, which regulate blood sugar levels.

Insulin: Insulin is a hormone secreted by the beta cells in a certain section of the pancreas. Because relatively small amounts of insulin are necessary in the body tissues, it is believed that insulin acts as a catalyst in cellular metabolism, i.e. the utilization of glucose in cells for energy.

Glucagon: Glucagon, like Insulin, is a pancreatic extract. Its primary function is the converting of hepatic glycogen to glucose. This produces an elevation of the glucose concentration in the blood.

OK, I know it sounds like just a bunch of dead words, which mean nothing to you; however the functions of insulin and the pancreas are vital to your health and quality of life. So please be patient, and learn as much as possible about their function. I'm giving you the Mayo Clinic's version (below) of how the Pancreas and Insulin work and how the disruption of the two can cause serious diseases, such as heart failure and diabetes. This is one version of the dilemma; there are others as Mayo later points out with other names and theories concerning the syndrome. However, they all agree that diet can control the various conditions which create the heart disorder, which is the leading cause of death in the United States, not to mention the largest growing disease: diabetes. To review the Metabolic Syndrome conditions, they are: increased blood pressure, elevated insulin level, excess body fat around the waist and abnormal cholesterol levels.

At the end of the Mayo report, we'll give you a breakdown of the good and bad foods to eat according to the Glycemic Index presented in a book entitled, The Glucose Revolution Life Plan, written by Jennie Brand-Miller, Johanna Burani and Kaye Foster-Powell. We will explain what the Index means and its' impact at that point in the book.

CHAPTER THIRTY
(1) METABOLIC SYNDROME

Definition: Metabolic syndrome is a cluster of conditions (as stated above) which occur together, increasing the risk of heart disease, stroke and diabetes.

Having just one of these conditions: elevated blood pressure, increased insulin level, excess body fat around the waist or abnormal cholesterol levels, isn't diagnosed as metabolic syndrome, but it does contribute to your risk as the syndrome suggests. If more than one of these conditions occur in combination, the risk becomes greater.

In reviewing your own personal conditions in reference to the metabolic syndrome, you have the opportunity to make aggressive lifestyle changes. Doing so can delay or derail the development of serious diseases resulting from the metabolic syndrome warning.

Research into the causes continues on in an effort to understand the underlying complexities, which link the group of conditions involved in the Metabolic Syndrome. As the name suggests, Metabolic Syndrome is associated with the body's metabolism, and likely to a condition called insulin resistance. Insulin (as stated earlier) is a hormone made by the pancreas, which aids in controlling the amount of sugar in the bloodstream.

The digestive system normally breaks down some of the foods eaten into sugar (glucose). The blood carries the glucose to body tissues, where the cells use it as fuel. Glucose enters the cells with the help of insulin. In people with insulin resistance, cells don't respond as they should to insulin, and glucose can't enter the cells properly. The body reacts by producing more insulin to help glucose get into the cells. The outcome is higher than regular levels of both insulin and glucose in the blood.

Even though the glucose elevation may not be considered high enough to qualify as diabetes, a raised glucose level still interferes with the body processes. Increased insulin raises the triglyceride level and other blood fat levels. It also interferes with the kidney function, resulting in higher blood pressure. These multiple effects of insulin resistance increase the risk of heart disease, stroke, diabetes and other conditions.

Insulin resistance remains a subject for

extensive research. There is a suspicion among scientists that genetic and environmental factors are probably involved with this condition; possibly predisposing certain families to the problem. Overweight and inactivity are also contributing factors.

What adds to the mystery of the condition is the disagreement among the experts: some don't even agree on the definition of metabolic syndrome or whether it even exists for that matter, as a distinct condition. Physicians have spoken about the various risk factors for years and identified them with different names, i.e. syndrome X and insulin resistance syndrome. Whatever they want to name it or define it as, still leaves the problem that this collection of risk factors is becoming more prevalent, leading to serious fatal diseases in humans.

The following factors increase the risk of developing metabolic syndrome:

...Age – The numbers indicate the older a person gets, the more prone he is to developing the metabolic syndrome: less than ten percent in their 20's, 40 percent in their 60s. Contrary to this, other research has shown about one in eight schoolchildren have three or more components of the syndrome. In league with this finding, other research has revealed an association between childhood metabolic syndrome and adult cardiovascular disease later on in life. The age research seems to be all over the board, but the facts contributing to the metabolic syndrome remain steady: if one fits into its' description, the results are fatal.

...Race – Hispanics and Asians are more commonly at a greater risk for metabolic syndrome than other races. Perhaps diet plays a role in their susceptibility to it.

...Obesity – The indicator used to determine obesity as a possible risk for metabolic syndrome is the body mass index (BMI): based on height and weight a percentage is calculated; results greater than 25% increases one's risk of metabolic syndrome. The positioning of body fat is also an indicator: abdominal obesity – having an apple shape is a higher risk rather than a pear shape.

...History of Diabetes – The likelihood of having the syndrome is linked to a family history of type two diabetes or a history of diabetes during pregnancy (gestational diabetes).

...Other diseases – High blood pressure, cardiovascular disease or polycystic ovary syndrome, which is a related type of metabolic problem affecting a woman's hormones and reproductive system, may also increase the risk of metabolic syndrome.

    Source: Mayo Clinic.com – Tools for a healthier lives, Mayo Clinic Staff, November 7, 2008, Page 1

CHAPTER THIRTY ONE
(2) THE GLYCEMIC INDEX

The Glycemic Index or G. I. ranks foods based on their immediate effect on blood sugar levels. Carbohydrate foods with the highest G. I. values, break down quickly during digestion, producing a fast blood sugar response. This puts too much demand on the pancreas causing over stimulation, where in time it may become exhausted, and genetically susceptible people may develop type 2 diabetes. This isn't all, too high insulin levels can be dangerous in other ways: where it is true our bodies need insulin for carbohydrate metabolism, too much of this hormone can have a profound effect on disease development, notably: heart disease and hypertension. Since insulin also influences the way we metabolize foods, it will dictate whether we burn fat or carbohydrate to meet our energy needs. This determines the storage of fat in our bodies. So the ideal is carbohydrate foods which breakdown slowly, releasing glucose gradually into the bloodstream, or have low G. I. values.

Here is how the Index works: by substituting

low G. I. foods (good) for high G. I. foods (bad), is all one has to do to put the theory into practice. This will eventually lower the dieter's overall glycemic index.

    Some of the best sources of carbohydrate include bread, crackers and baked goods, breakfast cereals, rice, pasta, potatoes and potato products. I'll repeat myself. Consuming low G. I. varieties of these foods will significantly lower the glycemic index of your diet.

High G.I. Food (bad)

Bread
Fluffy, light, smooth textured white
or whole wheat (made from
enriched wheat flour).
Rice:
Short grain sticky (Chinese or Italian),
jasmine.
Potatoes:
Instant mashed, red and white-skinned
Baking varieties.
Cereals:
Most processed cold breakfast cereals,
as well as quick and instant cooked
types, such as oatmeal
Crackers:
Most crackers (Saltines, Triscuits),
rice cakes
Fruit:
Mango, pineapple, dates, watermelon, raisins.

Low G.I. Alternatives

Bread:

Dense breads containing a lot of whole grains; sourdough and stone ground flour breads (types that don't contain any enriched flour)

Rice:

Long grain basmati, imported Japanese, Uncle Ben's converted, brown, long grain White

Potatoes:

All pastas, noodles, legumes, (including Soybeans, kidney beans, lentils, chickpeas, Baked beans), barley and bulgur (cracked wheat), sweet potato, yam, taro, new potatoes (have moderate Glycemic index).

Cereals:

Rolled oats, semolina, muesli, granola, All Bran with Extra Fiber Bran Buds, Grapenuts, Special K, oat bran

Crackers:

Ryvita, stone ground wheat thins, WASA

Fruit:

Apples, pears, citrus fruits, cherries, peaches, plums.

Source: The Glucose Revolution Life Plan, by Jennie Brand-Miller,
Ph. D., and Johanna Burani, M. S., R. D. , C. D. E., and Kay Foster-Powell,

B. Sc., M. Nutrition & Diet. Publisher: Marlow & Company, New York, NY, 2001, Page 11.

CHAPTER THIRTY TWO
OTHER FOOD INFORMATION

Smoked Meats:

Smoking cigarettes has been proven to be one of the leading cause of lung cancer. However, there is another kind of smoking, which scientists have found almost equally dangerous: smoking to preserve meat.

During the smoking process to preserve meat, millions of tiny carbon particles become lodged in the meat. They are small bits of carbon that resemble lamp black. Research has implicated this change in the meat, as a cause of stomach cancer.

Stomach cancer is on the decline in America, but it is still a prevalent health problem, and often fatal because it is difficult to detect it in its early stages. The decline in this killer may be related, in part at least, to a decline in the consumption of smoked foods.

Before refrigeration and freezers smoking meat was commonplace, as a means of preserving it in the warm weather. Today, freezing techniques allow a far safer form of meat preservation, and refrigerated transportation speeds fresh meat to markets daily. This has greatly reduced the need for smoked meat in the United States. Scientists suggest that this is why stomach cancer is declining in most areas, where smoked meat is not a staple item of the diet. Regardless, smoked sausage, fish, hams, bacon, and other products are still in widespread use in the country.

As a proof to support the researcher's conclusions, it has been shown that the stomach cancer rate goes up in the population areas that subsist mainly on smoked food. This research was concluded in a study conducted by Dr. Niels Dungal of Reykjavik, Iceland, and reported in the Journal of the American Medical Association. Dr. Dungal became interested in the stomach cancer problem, because the disease accounts for up to 45 percent of all cancers in Icelandic males. He concluded, " smoked fish contains large quantities of polycyclic hydrocarbons, known cancer causing agents."

Another report was issued concerning the link between smoked meat and cancer of the stomach. Dr. Charles C. Stock, biochemist in charge of experimental cancer chemotherapy at the famed Sloan-Kettering Institute, delivered his findings before the Montreal Medico-Chirurgical Society in Montreal, Canada. He supported Dr. Dungal's findings, confirming the incidence of stomach cancer as being extremely high in regions where meat and fish are smoked before consumption.

"Statistics show that in Japan alone 50 percent of cancer histories are stomach cases, whereas in the central United States, where fresh is plentiful, the number of stomach cancers is very low."

"In Iceland the number of stomach cancer cases is also very high," he said. "There is a very strong suspicion that this is in some way related to the large amount of smoked fish the natives eat."

He fed smoked fish to laboratory rats, and one third developed malignant tumors. Also he found a higher rate of stomach cancer in areas where people fish for their food.

Source: The Encyclopedia of Common Diseases by the Staff of Prevention Magazine, Rodale Press, Inc. 1976, page 322

CHAPTER THIRTY THREE

NITRITES

Unfortunately the cancer threat in commercial meats goes beyond whether or not they are smoked. The problem is the use of nitrates to color and preserve meats, as well as fertilizers for our crops. Nitrates in their original form aren't carcinogenic. However, under certain conditions, they can change composition to nitrites, which are cancer causers.

Usually, nitrite starts out as nitrate, a relatively harmless compound. Nitrates are nitrogen compounds essential to plant life. The bacteria in soil metabolizes organic nitrogen compounds, and releases nitrates. They in turn form the nitrogen, which can be absorbed and utilized by plants for growth. Certain plants, mostly greens such as spinach, accumulate relatively large but highly variable amounts of nitrate. Ordinarily, we wouldn't be eating enough of this natural nitrate to cause any problems. However in our modern technological age there is nothing ordinary about the food and water most of us consume.

In the natural state of organic farming,

plants get their nitrogen slowly as organic matter in the soil slowly decays. Today, in order to spur the maximum amount of growth from overworked soil, most farmers use large amounts of chemical fertilizer. The nitrogen in this fertilizer is already in the inorganic or nitrate form in which it can be immediately used by the plant. (1)The more of this fertilizer used, especially as the crop is ripening, the more nitrate winds up in our food and in our drinking water (Wolff and Wasserman, Science, July 7, 1972). (2)When vegetables are grown under shade, as they sometimes are in very hot areas, the nitrate count rises still higher, while the Vitamin C content diminishes! (Miss. Farm Research, July, 1962).

Also human sewage, estimated at millions of tons, winds up in rivers every day, further adding to the nitrate burden of drinking and irrigation water.

If this wasn't bad enough, there is another way in which nitrates enter our food. (3)Fully eight billion pounds of beef and pork, mostly in the form of hot dogs, luncheon meats and canned hams, are impregnated with nitrates and/or nitrites and eaten by Americans every year. (Robert Rust of Iowa State University, National Hog Farmer, July 22, 1972). The reason given for using this nitrate is to give aging and decomposing meat a healthy – pink or reddish color and perk up its' flavor. The need for such a lift in hot dogs, for example, becomes apparent, when you realize most hot dogs contain large amounts of added fat, cereal, water, and goat or old chicken meat. Nitrate allows this garbage to be kept in storage for weeks before the unsuspecting consumer buys it, and feeds it to his children.

    The argument from some processors is that the addition of nitrates retard bacterial growth and inhibits growth of the spores that cause botulism. Other processors point out that some nations have banned the use of nitrate without triggering an epidemic of food poisoning. As an alternative foreign processors often use natural preservatives or boil foods for longer periods of time prior to processing. In fact, one of the best safe preservatives is ascorbic acid or Vitamin C; this is the very factor, which protects against nitrate by destroying it.

    Note: This is what the experts thought about the use of nitrates. This isn't entirely true, as the following will bring to light.

Being profit driven, many meat processors rely on the most convenient method to keep their products looking fresh. What they discovered was, it wasn't the nitrate that keeps the color in meat. What takes place, is the bacteria normally found in meat changes the nitrate and reduces it to nitrite. It is the nitrite that does the cosmetic and preservative work. Disregarding the fact that nitrite is far more toxic than nitrate, many processors simply quit using nitrate and replaced it with nitrite, or used some nitrate along with it as a sort of nitrite time- release capsule. The label on the lunch meat may say 'sodium nitrate added', but you can be sure that by the time you eat it, much or all of the nitrate has been transformed into nitrite.

This is how nitrite works in the stomach: once there it combines with a substance called secondary amines to form compounds known as nitrosamines. Nitrosamines are known to cause cancer. All that is required for this reaction (called nitrosation) is the presence of the two in the stomach at the same time.

Source: The Encyclopedia of Common Diseases, by The Staff of Prevention Magazine, Published by Rodale Press, Inc., 1976, Page 323

(1) Wolf and Wasserman, Science July 7, 1972
(2) Miss. Farm Research, July, 1962
(3) Robert Rust of Iowa State University (National Hog Farmer), July 22, 1972

CHAPTER THIRTY FOUR

MORTALITY

15 Leading Causes of Death in the U. S., year 2000

Rank: Causes of Death

Number: All causes 2,404,624

1 Diseases of the heart (food related) 709,894

2 Cancer (food related) 551,833

3 Stroke (food related) 166,028

4 Chronic lower respiratory diseases 123,550

5 Accidents (unintentional) 93,592

    Motor vehicle 41,804

    All others 51,788

6 Diabetes mellitus (food related) 68,662

7 Pneumonia and influenza 67.024

8 Alzheimer's disease 49,044

9 Nephritis, nephritic syndrome and nephritis 37,672

10 Septicemia   31,613

11 Suicide 28,332

12 Chronic liver disease and cirrhosis (food related)   26,219

13 Hypertension and hypertensive renal disease 17,964

14 Pneumonitis due to solids and liquids 16,659

15 Homicide 16,137

   All others 400,401

Source: U. S. National Center for Health Statistics, Department of Health and Human Services.   (The World Almanac and Book of Facts, 1999)

CHAPTER THIRTY FIVE
DEPRESSION

Note: Isn't it odd why depression should be so overwhelmingly present in the United States, the land of opportunity and liberty and money. You would think everybody would be happy here, wouldn't you? But, the fact is they're not.
An estimated 33 to 35 million adults in the United States are likely to experience depression at some point during their lifetime. The disease affects men and women of all ages, races, and economic levels. However, women are at a significantly greater risk than men to develop major depression. Studies show that episodes of depression occur twice as frequently in women as in men.

Although anyone can develop depression, some types of depression, including major depression, seem to run in families. Regardless whether depression is genetic or not, the disorder is believed to be associated with chemical level changes in the brain such as serotonin and nor-epinephrine.

Definition: There are two reasons or classifications for depression: reactive (exogenous) depression and endogenous depression. Reactive depressions are often caused by a person's response to a loss of pleasure or interest in activities and everyday living (i.e. loss of a loved one, a debilitating illness, not meeting one's expectations or loss of a job, pet, friend, etc.). This is normal depression, which generally remits in several months without the use of medications. The mobilization of support systems and, in certain conditions psychotherapy, are useful in dealing with exogenous or reactive depression.

Endogenous depressions are characterized by the absence of external causes for depression. This type of depression may be brought on by genetic determination and biochemical alterations (1) Csernansky and Hollister, 1982. Antidepressant medications are useful in the treatment of this type of depression.

The current classification for depressive disorders has eliminated the use of the above terminology (2) Andreason and others, 1980. Under their classification, major affective disorders are defined as bipolar disorders or mixed type and mania.   This also includes major depression (single episode or recurrent episodes), along with atypical affective disorders, which include depression.   Psychiatrists have debated over whether this new classification is an improvement over the previous (endogenous, exogenous, or manic depression) types of classification.   They argue it is important for the clinician to have a diagnostic framework from which to work.   So, if the experts or well educated professionals in the field disagree, where does that leave you and me with our depression problems?   My colleague and I feel you have the answer to that question inside yourself.   After all it's your mind.   Later on in the book, we will discuss possible solutions which are within the average person to help with or conquer the depression disease.   For now let us return to "definition."

Criteria for major depression are anxiety, crying spells, guilt feelings, self-pity, pessimism, loss of interest in life and social activities, psychological symptoms (low self-esteem, poor concentration, hopeless or helpless feelings, suicidal or increased focus on death), physiologic manifestations (sleep disturbances which may range from insomnia to hypersomnia, decreased interest in sex, complaints of fatigue, loss of energy, menstrual dysfunction, headaches, palpitations, constipation, loss of appetite, and weight loss or weight gain), and thinking alterations (a decrease in concentration or attention span, complaints of poor memory, confusion, delusions relating to health, persecution, or religion, and hallucinations if the client is also psychotic). Mood variations are often worse in the morning. Any of this sound familiar?

Atypical depressions usually are often of briefer duration and not as severe. They are non-responsive to the tri-cyclic antidepressants. Mood changes are usually worse in the evening: panic attacks, phobias such as agoraphobia, and physiologic complaints are often present then. Again (1) Csernansky and Hollister (1982) and others believe certain inhibitors are the drugs of choice for this type of depression. Other measures to treat depression include electroshock therapy, psychotherapy, reduction of environmental stressors, and milieu therapy. Antidepressant drug therapy in combination with one or more of the above adjunct measures can be more effective than drug therapy alone.

Source: Mosby's Pharmacology in Nursing, by Leda M. McKendry, Ph. D, R. N. and Evelyn Salerno, Pharm. D., R.PH., Published by Mosby – Year Book, Inc., 1992 – Page 337

(1)Csernansky and Hollister, 1982

(2)Andreason and others, 1980

Author's Note: Now that we know what depression is or what the experts think it is, let's examine it further in regard to everyday living and the environment. The worse problem we all have with these two related subjects is getting along with each other. Americans, I'm sorry to say, appear to hate one another. In my studies on the subject, I find my fellow countrymen are rude, violent, selfish, polluting and uncaring. This entire book proves that fact.

Yes we hate one another! It is rare for me to go anywhere, where people gather without experiencing this dislike. There are signs posted on well –traveled highways warning motorists:

beware of aggressive drivers," "slow down – save a life," "buckle up – it's the law," "don't tailgate," and the one that stands out in my mind the most, "We have many children, but none to spare – please drive carefully."

Signs are expensive; for the state to budget money for this project means they have good reasons to do so: they know the statistics. To me this is proof that the enemy is out there waiting to hurt us. And, unfortunately, he does (refer to the above death statistics on motor accidents). Most of the collisions are caused by aggressive drivers: drunk or under a drug influence, a mental disorder, immaturity, or not paying attention to the highway and ignoring the rules governing it.

American's favorite car isn't a car; it is a pick-up truck with a box on the back instead of a loading bed. This is known as the SUV or Sport Utility Van. What sport do they intend to use it for: bullying other motorists, who drive more gas efficient, smaller cars? Although this must be a lot of fun for these fools, it isn't really funny at all, because these vehicles are involved in more collisions than their smaller counterparts i.e. about

10% higher. Granted the SUV stands up better in a minor fender bender, but in a more dynamic accident on a major highway their record is more serious and fatal.

The cell phone is indeed an excellent invention, but in a moving vehicle it increases the possibility of a car collision. It is against the law in most states to use the cell phone while driving, but alas, the laws are badly enforced by local police. Do you know anybody who ever got ticketed for using a cell phone, while operating their car? I guess it is just one of those laws that no one obeys, like going through a 'yield' sign, and the police don't take it seriously either.

So you get to your destination in one piece. If you had rudeness on the road, you will surely receive the same, when you arrive at the mall or shopping place or whatever. Why is it so hard to say 'excuse me' when you bump into someone or move ahead of them in a crowded place? Not hard at all, but with many Americans it is, since they haven't the time for this easy to say nicety, or so they tell themselves. Doing a simple errand, one is bound to run into different rudeness of this nature. All this for what reason; what does it take to practice a little courtesy: an awareness and a determination to make being nice a part of your life. How hard can that be?   It shouldn't be, but to some apparently it is, because rudeness persists.

True the above lack of manners are minor depressions, but they are frequent and can affect ones' overall frame of mind. Add that to the major depressions, I'm about to list and you come up with depression on a mass scale, which we have in the United States. The following is that list, which I define as, More Depression.

CHAPTER THIRTY SIX
MORE DEPRESSION:

1. More Depression – The murder rate in the United States is 200 times greater than in Japan, where no private citizen can legally buy a handgun. We are talking handguns here, not rifles; this weapon is designed and marketed for the killing of human beings. In hunting animals it is ineffective.

The United States has the world's highest death rate per capita from firearms, a new gun is sold every 13.5 seconds. The sad truth is that almost twice as many Americans were killed on the 'home-front' by firearms in the Vietnam War decade of 1963-73, than were killed in the war effort in that country. The figures, respectively, were 84,633 and 46,752.

Source: (1) Isaac Asimov's Book of Facts, Bell Publishing Company, New York, NY, by Red Dembner Enterprises Corp., 1981 Edition, Page 225

2. More depression – 1865 – April 14, United States President, Abraham Lincoln shot in

Washington D. C.; died April 15. Known for freeing the slaves and preserving the Union.

1963 – United States President, John F. Kennedy, shot in Dallas, TX. Known for championing the rights of the average American.

1968 – April 4, Rev. Dr. Martin Luther King, Jr. fatally shot in Memphis, Tenn. Known for advancing the rights of Black Americans.

1968 – June 5, Sen. Robert F. Kennedy (D, NY) shot to death in Los Angeles, CA. Shared the same beliefs as his brother John F. Kennedy.

1980 – John Lennon, former Beatle and peace activist was shot and killed December 8 in New York City.

Note: And yet the American leadership refuses to pass serious gun legislation. Citing the U. S. Constitution's Second Amendment, Right to keep and bear arms. "A well-regulated Militia, being necessary to the security of a free State, the right of the people to keep and bear arms, shall not be infringed."

Here is another right – the right to walk the city's streets safely. Total deaths by firearms, in the year 1999 – 28,874.

Source:

Centers for Disease Control & Prevention

3. More depression – Freedom of religion – United States Constitution – First Amendment: The Bill of Rights, Religious establishment prohibited. Freedom of speech, of press, right to assemble and to petition. "Congress shall make no law respecting an establishment of religion, or prohibiting the free exercise thereof; or abridging the freedom of speech, or of the press; or the right of the people peaceably to assemble, and to petition the Government for a redress of grievances.

Landmark Decision of the U. S. Supreme Court: 1962: Engel vs. Vitale. The Court ruled that public school officials could not require pupils to recite a state-composed prayer, even if the prayer was nondenominational and voluntary, because this would be an unconstitutional attempt to establish religion.

Comment concerning free speech and religious freedom:

The Court has now made the following documents illegal in light of the Engel vs. Vitale ruling:

...The Star-Spangled Banner, by Francis Scott Key, which is our country's national anthem, "...And this be our motto: '"In God is our trust!'".

...The Pledge of Allegiance to the Flag, quote "I pledge allegiance to the flag of the United States of America, and to the Republic for which it stands, one Nation under God, individual, with liberty and justice for all." (The phrase 'under God' was added to the pledge on 6/14/54.)

...Lincoln's Gettysburg Address: quote, "... that this nation under God shall have a new birth of freedom, and that government of the people, for the people shall not perish from the earth."

...Presidential Oath of Office: "I do solemnly swear that I will execute the office of President of the United States and will, to the best of my ability, preserve, protect, and defend the Constitution of the United States, so help me God." (The President Elect's right hand on the Holy Bible.)

...And of course all forms of minted money, which carry the statement, "In God we trust."

In addition, Christmas decorations (Freedom of Religion?) aren't permitted in Post Offices or any other government buildings, unless paid for by the employees. How does this go against Church and State separation? Don't you think the Court is stretching this a bit?

A suggestion to the President of the United States: as a checks and balance measure, write an executive order telling all our government people they can have their Christmas trees, and pray all they want in the buildings where they work. After all we are guaranteed freedom of religion in the same Amendment, as well as freedom of speech. By the Court saying Americans can't pray here, but they can pray over there is the same as saying, they don't have freedom of religion at all. The Court can't have it both ways.

4. More Depression – Due Process of Law, Amendment Five, Bill of Rights

"No person shall be held to answer for a capital, or otherwise infamous crime, unless on a presentment or indictment of a Grand Jury, etc.....nor shall be compelled in any criminal case to be a witness against himself, nor be deprived of life, liberty, or property, without due process of law, etc."

Supreme Court's Interpretation: 1966:

Miranda vs. Arizona.

The Court ruled that, under the guarantee of due process, suspects in custody, before being questioned, must be informed that they have the right to remain silent, that anything they say may be used against them, and that they have the right to counsel.

Comments: By the way, Miranda, a cheap criminal with a rap sheet as long as his arm, was set free because of this ruling. How many other criminals were set free because of this decree?

One other point, if the suspect is afforded this right, shouldn't his victim(s) be given the same privilege? Shouldn't their rights be explained to them, since most private citizens aren't familiar with the law, and often believe the law is on their side. Isn't that what Justice is: the balance of the scales; what is provided for the one side is equally provided for the other?

5. More Depression – Noise: On a sunny day, you want to relax and sit outside on a lawn chair, sunning yourself and perhaps reading a book. However your next door neighbor decides he is going to play his radio; not with ear phones on, but for everyone, including you to hear,

whether you like it or not.

Second scenario, you love your garden. Another sunny day, you want to go out in your back yard and attend to your roses, azaleas, water lilies, or whatever. The next door neighbor decides to put Rover, their pet noise maker, out in their yard, who immediately takes a dislike to you, and continues to bark and bark and bark. When you complain to your neighbor about it, he tells you that the barking is your fault, since you don't know how to handle dogs.

Third scenario, the weather is starting to turn warm, so you open your bedroom window. You were out late last night, and want to sleep in. Your next door neighbor has different plans for you. Right under your bedroom window, he decides to blow his power blower, until every dead leaf is removed; a rake simply won't do the job.

What rights do you have with these invasions of your privacy with noise?
Unless it is extremely excessive, and not on their property, you probably have no rights at all from your local authorities. Is this fair? Not to me it isn't!

Here is a Supreme Court Judgment on

Noise pollution (not U. S.). Their Court rendered the verdict November 9, 2007 concerning the right of one party to use loudspeakers against the annoyance, disturbance and harm caused to those compelled, against their will, to listen to the amplified sounds, emanating from their loudspeakers. Basically their Court held, nobody can claim the fundamental right to create noise by amplifying the sound of his speech. Consequently, just as one has the right of speech, others have the right to listen or decline to listen. Additionally, their Court reasoned a great deal of harm is done to school children, whose studies are disturbed and to the sick, who are recovering from illness. It noted, noise can produce serious physical and psychological stress.

OK I know this isn't U. S. law, but couldn't it be? If the Court can protect cheap criminals, couldn't it equally protect hard working honest citizens from unnecessary noise in their own homes? Don't you think it would be a good place to start in correcting noise pollution? Why would it be that hard to enforce nationwide? Put the police to work.

After all an amplified radio never put a roof

on a house, fixed faulty electricity or plumbing or any needed home repairs. It is illegal on an airplane, train, bus and movie house. Why shouldn't it be illegal where a person lives – his home? What is more important than one's home?

Source: November 9, 2007, Supreme Court Judgment On Noise Pollution (not the U. S.), Citizens' Movement for Good Governance (CIMOGG) Articles and Publications.

CHAPTER THIRTY SEVEN

THE MEDICAL/HEALTH CARE INSURANCE COMPLEX

Note: This is actually part of More Depression, but since it is so long I decided it deserved a chapter of its' own.

Up to this point I've ascertained our American society is getting its' population gradually sick by polluting the air we breathe, destroying the earth's ozone protection and poisoning the food we eat with preservatives and artificial coloring, together with the improper disposal of radioactive wastes. When we eventually are stricken with a terrible ailment such as cancer, heart disease, or diabetes, etc. from this, we are at the mercy of the Medical/Health Care Insurance Complex, which dictates the kind of health care the patient will receive. Briefly, this Complex responds to the patient by how much money he or she has put into it. Rule of thumb: the more money put into the insurance policy by the holder, the better the care provided i.e. doctors, hospital services, preventive medicine, etc.

In doing my research and gathering this data, one disturbing fact stands out: the United States spends a great deal more money per capita on health care than other developed nations. I've devoted this chapter to trying to understanding why. My partner and I didn't want to assume that a good many people are getting rich through the system. However the facts are there, and this is what we have presented. Getting rich on sick people! Here is what we found out.

1. The United States is the only wealthy and industrialized nation that does not have universal health care. Here are some of the figures: 85% of our citizens have some form of health insurance; either through their employer (59%), purchased individually (9%), or provided by government programs (28%). Americans without health insurance coverage during 2007 totaled about 15% of the population, or 46 million people. For some reasons, which aren't apparently clear, the cost of health care insurance is rising faster than wages or inflation.

2. Emergency Medical Treatment and Active Labor Act (EMTALA), enacted by the federal government in 1986, requires that hospital

emergency rooms treat emergency conditions of patients regardless of their ability to pay. It is the only medical care for America's uninsured. The problem with this is the government hasn't fully compensated public and private hospitals for the full cost of care mandated by EMTALA. To be exact more than half of all emergency care service in the country goes uncompensated. This unfunded mandate has led to financial pressures on hospitals in the past 20 years, causing them to consolidate and close facilities. What is most disturbing, this has contributed to emergency room overcrowding. When you're bleeding to death, waiting around for a doctor doesn't do it for you.

According to the Institute of Medicine, emergency room visits in the U. S. grew by 26%, between 1993 and 2003. To make this statistic worse, during the same period, the number of emergency departments declined by 425. Hospitals, in an attempt to improve their cash flow and become more solvent, tried billing uninsured patients directly under the fee-for-service model. But this is like getting blood out of a tomato; it just isn't there. Most ER treated people simply can not pay their bills and escape into bankruptcy, when hospitals seek legal action against them. Although fee-for-service model can work in some sectors, in the case of emergency care, it doesn't allow the patient to shop for a better price, nor does it permit price negotiations with the hospital.

Mentally ill patients are a special challenge for hospital emergency rooms. EMTALA regulations dictate that mentally ill patients, who enter emergency rooms, are to be evaluated for emergency medical conditions. Once a mentally ill patient becomes stable, a regional mental health agency is notified for their evaluation. The agency determines if the patient is a danger to himself or others. If he is, then arrangements are made to admit him to a mental health facility for further evaluation by a psychiatrist. Mentally ill patients can be held for up to 72 hours, after which a court order is required.

3. Price of Prescription Drugs –

During the 1990s, the price of prescription drugs in America has 'gone through the roof' so to speak. Many citizens discovered that neither the government nor their insurer would cover the cost of such drugs. The patient was on his own with these outrageous price increases from large pharmaceutical companies. In absolute currency, the United States spends the most on pharmaceuticals per capita in the world. Why?

Here is what the U. S. government has to say about it (through the Office of the United States Trade Representative). American drug prices are rising because our consumers are effectively subsidizing costs, which drug companies can't recover from consumers in other countries (because many other countries use their bulk-purchasing power to aggressively negotiate drug prices). This rhetoric is consistent with the primary lobbying position of the Pharmaceutical Research and Manufacturers of America. They feel that the governments of such countries are free riding on the backs of American consumers. These countries should either deregulate their markets, or raise their domestic taxes in order to compensate U. S. consumers by directly remitting the difference to the drug companies or to the U. S. government. In so doing, pharmaceutical companies would be able to continue to produce innovative pharmaceuticals, while lowering prices for U. S. consumers. (Here I'm going to pause for a moment before going on to chuckle a bit. Forgive me, I always laugh at things that aren't funny.)

OK, I feel better now.

Currently, America, as a buyer of pharmaceuticals, negotiates some drug prices, but is forbidden by law from negotiating drug prices for the Medicare program. The law in question is a Medicare Bill passed by the Republican controlled 109th Congress, and signed into law by President George W. Bush in 2005. Democrats have charged that the purpose of this provision is merely to allow the pharmaceutical industry to profiteer off the Medicare program, which is already in imminent danger of becoming financially insolvent. Could be, they are right.

4. Delays in Seeking Care and Increased Use of Emergency Care:

Americans without health care insurance don't usually get regular checkups by a doctor, nor use preventive medical services. They are the victims of the system, since the sooner an illness is diagnosed the more likely it can be treated inexpensively, and of course successfully. When these people seek care, the illness has developed into a medical crisis, which is more expensive to deal with than ongoing outpatient treatment. A study published in JAMA (Journal of American Medical Association) recently concurred with this, and the fact that uninsured people were less likely than the insured to receive any medical care after an accidental injury. The same holds true for the uninsured with an injury, who are twice as unlikely to go for the recommended follow-up care. On the other hand the uninsured are twice as likely to visit hospital ER's with an ordinary illness, burdening a system meant for true emergency needs.

Researchers with the American Cancer Society in a recent study found that people, who lacked private insurance, were more prone to be diagnosed with late-stage cancer than the insured. This was true of both the uninsured as well as those covered by Medicaid. It was concluded in the study, individuals without private insurance were not receiving the proper care in terms of cancer screening or timely diagnosis and follow-up with health care providers.

The dollar and cents fact remains, the cost of treating the uninsured must often be absorbed by the medical providers as charity care; passed on to the insured through higher health insurance premiums, and paid by taxpayers through higher taxes. No such thing as a free lunch.

Source (points 1 through 4: Health Care in the United States – Wikipedia, The Free Encyclopedia, Page 9

5. Getting down to the nuts and bolts of the chapter: How the U. S Health Care Spending Compares To Other Developed Countries:

The United States Is high, very high. Here is what the Organization for Economic Co-operation and Development (OECD) compiled, comparing health care expenditures for countries with above-average per capita national income. As they say in Poker: read them and weep.

Australia $3,128, Austria $3418, Canada $3161, Denmark $2,972,

France $3,191, Japan $2,358, etc.

Compared to the United States $6,037

Despite this relatively high level of spending, the United States does not appear to achieve substantially better health benchmarks compared to other developed countries.

Source: OECD Health data 2007 from the

OECD Internet subscription database updated July, 2007. Copyright OECD 2007, Page 4

6. Why Is American Health Care Spending So High?

1. Higher prices for the same health care goods and services than are paid in other countries for the same goods and services. Reasons: EMTALA laws, unregulated pharmaceutical companies price gouging, high malpractice insurance for physicians, etc.

2. Significantly higher administrative overhead costs than are incurred in other countries with simpler health-insurance systems.

Note: At parties or whenever I'm in a crowd, I always like to tell the story of when I had an operation in the early 60's. All I had to do was present my Blue Cross cards to the admittance clerk at the hospital, and every payment was taken care of; all I had to do was sign my name to the forms. The health care insurance was part of a benefits package, where I worked (paid for by the company). What has happened in this country, where 45 million of us can't get the proper medical treatment which I had when I was a young man? If we did it then, why can't we do it today?

Oh, by the way, the operation was a success.

3. There is more widespread use of high-cost, high-tech equipment and procedures than are used in other countries. Yes that is true, but other countries are absorbing this cost too. If it improves a patient's chances for getting cured, it is money well spent. This is the question that must be asked: is it money well spent? Making somebody rich, who doesn't do anything to deserve it, isn't money well spent!

4. Higher treatment costs triggered by our uniquely American tort laws, which in their context of medicine can lead to "defensive medicine" – that is, the application of tests and procedures mainly as a defense against possible malpractice litigation, rather than as a clinical imperative.

As a side effect of this, the American physician must pay extremely high premiums for malpractice insurance. Guess who he passes this expense on to.

Source: Why Does U. S. Health Care Cost So Much? – Economic Blog –

NY Times, Excerpts from the book, Why Does U. S. Health Care Cost So Much? By Uwe E. Reinhardt, Economist at Princeton University. December 24, 2008, Pages 1 to 13.

End of: MORE DEPRESSION

CHAPTER THIRTY EIGHT
DEPRESSION FOR NO REASON

I'm separating Depression For No Reason from Depression for this is different, even though it feels the same for its victim. Under the previous written definition of depression, it was categorized as endogenous depression or the absence of external causes for depression. This depression is so puzzling, that we felt it necessary to expand and write more on the subject, giving it a chapter of its own. Bear with us if we duplicated some of the information.

Why would anybody be depressed for no reason, since there is so much to be depressed about, without being depressed for no apparent reason? I'm adding the word 'apparent' because it is the key to understanding this riddle. There is a reason, because there is a reason for everything, and science is beginning to involve itself in research to expose this mystery. Knowing that there is a problem is the first step in solving it. Some ideas have come to light, and I would like to share them with you, characterized with the different disorders.

1... (1)BIPOLAR DISORDER – This affliction is an illness, which causes extreme mood changes from manic episodes of high energy to the extreme lows of depression. The name given to it is manic-depressive disorder.

It can cause behavior so extreme that the person can't function at work, in family or social situations, or in relationships with others. People with bipolar disorder can feel helpless and hopeless to the point, where they contemplate suicide.

Having this disorder isn't totally hopeless because these people aren't alone. Talking with others, who suffer from it may help the victim learn that there is a possibility for them to find a better life. Treatment is available, which can help getting them back in control.

The cause of bipolar isn't completely understood. One fact is, it runs in families. Also it is affected by one's living environment or family situation. A possible cause is an imbalance of chemicals in the brain.

The symptoms depend on the person's mood swings. In a manic episode, the victim may feel happy, energetic, or on the edge, like he needs very little sleep, and is overly self-confident. Some spend a lot of money or get involved in dangerous activities, when they are manic.

(Note: Allow me to interject here concerning the 'dangerous activities' so mentioned. Could it be that these people are driving automobiles, and if so, should they be?)

After a manic episode, the person may return to normal, or have a mood swing in the opposite direction to feelings of sadness, depression, and hopelessness. When depressed he may have trouble thinking and making decisions; have memory problems; lose interest in things he enjoyed in the past. He may even have thoughts about killing himself.

Treatment for bipolar should begin immediately upon diagnosing the disorder. In dealing with a manic episode it is important to recognize the early warning signs, so that immediate medication can be administered, which can treat the manic phases. Various medicines are used to treat bipolar disorder. The treatment becomes a trial and error procedure before the patient finds the right combination of medicines that works. Most bipolar people need to take a medicine called a mood stabilizer every day. Antidepressants should be used carefully for episodes of depression, since they cause some patients to move into a manic phase(s). Regular checkups are important, so that your doctor can tell if the treatment is working. Counseling for the patient and his family is also a vital part of the treatment. It can help the patient cope with some of the work and relationship issues, which may have caused the illness.

> (1) Source: Health – Vital Information with a Human Touch, Copyright 2008 Health Media Ventures, Inc.

2. (2) GENERALIZED ANXIETY DISORDER – or GAD is characterized by excessive, exaggerated anxiety and worry about daily life events. People with symptoms of the disorder often expect disaster and worry constantly about neighbors, health, money, family, work or school. The GAD victim's concern is often unrealistic or out of proportion for the situation. Everyday life becomes a constant state of worry, fear and dread. Eventually, the anxiety is so dominant that it is all consuming for the afflicted. About four million adult Americans suffer from this disease during the

course of a year. It often begins in childhood or adolescence, but adults aren't exempt from it either. Women are more prone to it than men.

Some of the symptoms are: excessive, ongoing worry and tension; an unrealistic view of problems: restlessness or a feeling of being edgy; irritability; muscle tension; headaches; sweating; difficult concentrating; nausea; the need to go to the bathroom frequently; tiredness; trouble falling or staying asleep; trembling; being easily startled.

The cause of Generalized Anxiety Disorder is not fully known, but a number of factors, including genetics, brain chemistry and environmental stresses, appear to contribute to it.

A. Genetics – Some researchers suggest that GAD is passed on in families, increasing the likelihood that a member of such a family is more prone to inherit the disorder than in others.

B. A brain chemistry abnormality has also been associated with GAD. The theory brings up the possibility that a GAD victim may have disproportionate levels of certain neurotransmitters in the brain. These chemicals work as special messengers, moving information from nerve cell to nerve cell. If they are out of balance, messages can't get through the brain properly. This can alter the way it reacts to certain situations, leading to anxiety.

C. Environmental factors, such as trauma and stressful events, i.e. abuse, the death of a loved one, divorce, changing jobs or schools, may lead to GAD. The disorder also may become worse during periods of stress. The withdrawal from addictive substances, including alcohol, caffeine and nicotine, can also worsen anxiety.

Note: I realize that 'C' is a Reactive depression. After all we don't live in shell. The two depression are sometimes related.

Medical treatment – See your doctor. However, most treatments deal with the symptoms and are often temporary. I'll have more to say on this towards the end of the sections.

(2) Source: Anxiety Symptoms, Causes, Types, Signs and Treatment on Medicine Net.com. 1996 – 2008 Medicine Net, Inc. All rights reserved.

3. (3) SUICIDAL DEPRESSION – Probably the most serious result of depression is the risk of suicide. The deep despair and hopelessness that

goes along with the disorder can make the afflicted individual conclude that self-destruction is the only way to make the pain go away. The suicidal individual often gives warning signs of their intentions. The best way to prevent suicide is to understand these warning signs, when you see them in a friend or relative. If you conclude that the individual is suicidal, it is pertinent to get involved by pointing out the alternatives, showing that you care, and getting a professional involved.

4... POSTPARTUM DEPRESSION – Strong emotions often occur in the mother after she gives birth: in fact it is considered normal. She is recovering from an extremely difficult and painful experience. Most likely she is sleep deprived, and adjusting to the responsibilities of parenthood. Postpartum depression, is longer lasting and more serious than many realize. What can be especially upsetting to mothers suffering from the depression are feelings of wanting to avoid or even harm the baby. Postpartum depression doesn't have to happen immediately. It can occur up to a year after childbirth.

5... ALCOHOL AND DRUGS can cause strong depression symptoms on their own. They make the user more venerable to depression, even if he decides to stop using them. Some people try to treat themselves with alcohol and drugs to self medicate, but this only worsens the problem.

Source: (3) Depression – Signs, Symptoms, Types and Ways to Get Help Joanna Saisan, MSW, Melinda Smith, M. A., Robert Segal, M.A. and Jeanne Segal, Ph. D, contributed to this article. Last modified in July, 2008

SOME HELFUL SOLUTIONS:

1. Depression Recovery – The answers to depression recovery are just as vague and elusive as the disorder itself. No one treatment is appropriate in all cases. If you or a loved one possess the symptoms of depression, take time to explore the many treatments available. The best approach could be a combination of self-help strategies, lifestyle changes and professional help. Start small in your recovery strategy, and ask for help. A strong support system in place will speed your recovery. Taking this on by yourself often fuels depression, so reach out to others, even when you feel like being alone. Go to family and

friends with what you're going through and how they can help.

2. Healthy Lifestyle Changes – Lifestyle changes are easier said than done, but they can have a big impact on depression. Ask yourself, what changes could I make to support depression recovery? Some self-help strategies that can be effective are:

..Pick and choose supportive relationships;

..Regular exercise and proper sleep;

..Maintain a healthy, mood-boosting diet;

..Control your stress by practicing relaxation techniques;

..Avoid negative thought patterns.

3. Medication may relieve some of the symptoms of moderate and severe depression, but it isn't a cure nor a long-term solution, since it doesn't relieve the underlying problem. There are antidepressant drugs on the market, whose manufacturers imply that chemical imbalances in the brain cause depression and their medication can correct this imbalance. The answer is not that simple. Depression involves more than 'bad' brain chemistry. Successful treatment of depression, looks at every aspect of the patient's

life, and makes any necessary changes, which will make him happy.

One other piece of important information, antidepressant medications may cause side effects and safety concerns. Also, getting off them can be very difficult. If you're wondering if an antidepressant medication is right for you. First learn all the facts about them, their effects and eventual withdrawal. Consult your physician with your findings. It may take time and some trial and error, but the effort to determine their proper use, could bring satisfactory results.

Source: Depression – Signs, Symptoms, Types and Ways to Get Help Joanna Saisan, MSW, Melinda Smith, M. A., Robert Segal, M. A., and Jeanne Segal, Ph.D., contributed to this article. Last modified in July, 2008.

CHAPTER THIRTY NINE

ADDICTION

Addiction is a topic for a book on its own. For this book, I'll abbreviate the subject to give you an overview and a warning of how serious addiction is, and how it affects lives. Of course, lives affect the environment.

First off, what is addiction? Stedman's Concise Medical Dictionary gives the following meaning: "...habitual psychological and physiological dependence on a substance or practice which is beyond voluntary control."

Source: Stedman's Concise Medical Dictionary, Williams & Wilkins, Second Edition, 1994, Page 18

This is a short but concise meaning to a big problem. The disease affects millions of Americans. The substances of addiction, which I'll list the more common ones, are not only addictive but carcinogenic as well, i.e. cancer causers, it gets you both ways, in the mind and the body and opens you up for disease.

Note: The exception to this is table wine (taken in moderation), which has anti-carcinogenic properties, i.e. the darker red the wine is, the better.

What type of person is likely to become addicted? The question is a good one in light of the fact that most people aren't addicted. Why? We have a theory, which we would like to share with you.

When a baby is born he or she comes into this world either strong willed or weak. The strong are positive and self-assured: they are capable of thinking on their own and most importantly, knowing the difference between right and wrong. Knowing what's right is in line with a 'doing good' instinct inherent in all human beings. (Yes, we know nobody's perfect; even the strong make mistakes, but they realize them and return to the path of good.)

The good instinct is so strong that it dominates in all human beings, and never goes away. Equally inherent is a secondary instinct called selfishness. The strong person doesn't give into this; the weak do, and their lives are concentrated around this drive in them. As a result they aren't in line with their prime instinct: knowing the difference between right and wrong and practicing right accordingly. Consequently, the selfishness in them eventually leads to severe mental pain and anguish.

    Let me explain in more detail, why doing right is man's main directive. First off, it is a survival mechanism. We feel it has brought man to the position he now occupies in the world, since right is constructive, not destructive. It has guided humans throughout their existence. The drive is as important to man's survival as food and water. Going against it, and giving way to selfishness, is contrary to man's nature, and to his survival responsibility. How does this happen? The selfish or weak person is afraid and unsure of himself. The fear insists that their ego and desires be fulfilled at any cost, even crossing over the line to doing wrong. This causes depression, because they are going against the human's main instinct of doing right. To appease the pain they turn to self-medicating (drugs & alcohol), which lead to addiction. In the next chapter entitled Alcoholics Anonymous, we will review Bill Wilson (AA's founder) twelve steps to recovery.

I believe his findings back up our theory: the formula – concern for others and doing right equals recovery.

Our warning to parents of selfish children is, don't feed or encourage such behavior. You can have an impact on your off spring's selfishness by simply telling them, "No," when they want to do something wrong. That is assuming that you, the parents, know the difference. Find out, if you don't. The information is contained in a book called, The Holy Bible. Do it while they are still young. Waiting until they are teenagers is often too late!

Drug addiction places an enormous burden on society through its repercussion on the crime rate and healthcare. The economic costs of addiction have been estimated at 80 billion dollars a year in the United States alone. Many western countries have invested heavily in research toward understanding, treating and preventing addiction. And yet the crisis continues and worsens with little advancement on a cure. It is a killer in more ways than one. My belief is that there isn't a family in the United States, which hasn't been affected by it in some way: It eats at the very core of the inflicted individual and everyone around him.

Common Drugs of Addiction:

1. Amphetamines and Cocaine are stimulant drugs. They work directly on the monoamine (especially dopamine) retransmitted systems producing a euphoria, when taken intravenously.

(1) Amphetamines: heavy users may develop psychosis characterized by aggression, delusions of persecution, depression, paranoia, euphoria, and fully formed visual and auditory hallucinations.

(2)Cocaine: heavy users display restlessness, hypertension, hallucinations, nausea, vomiting, and muscle spasms, which could lead to respiratory failure, convulsions, coma, and circulatory collapse.

The effects of withdrawal include mild depression, fatigue, sleep disturbances, increased appetite and anxiety.

2. Opiates are analgesics, which include morphine and heroin (deactivated morphine). The drugs are usually injected. Many of their subjective effects include a sense of well-being and euphoria. They act through opiate receptors

of the mu type, to which naturally occurring chemical messengers such as B-endorphin and the enkephalins also bind. Withdrawal symptoms include mild depression, nausea, muscle cramps, tear production, diarrhea, sweating, anxiety and fever, i.e. cold turkey.

3. Alcohol acts in many ways, similar to those anxiety-relieving drugs such as the benzodiazepines (abuse of which can also lead to addiction). Withdrawal symptoms include autonomic hyper-reactivity, nausea, hand tremor, anxiety and hallucinations. Alcohol has strong sedative effects, and may lead to memory loss. Most importantly, it can render an automobile driver unable to safely operate his vehicle.

4. Nicotine affects receptors for the neurotransmitter Acetylcholine, found in the neocortex, hippocampus and midbrain. Withdrawal symptoms include mild depression, insomnia, anxiety, restlessness, decreased heart rate and weight gain.

(2) Burning tobacco can generate gaseous substances: included are carbon monoxide, carbon dioxide, hydrogen cyanide, ammonia,

volatile nitrosamines, and others. Smoking has been identified to be a known carcinogen, causing the following disorders: cancer of the bladder, lung, buccal cavity, esophagus, and pancreas. Other illnesses are: pulmonary emphysema, chronic bronchitis, coronary heart disease, and myocardial infarction.

5. Cannabis (the dried flowering spikes of the hemp pistillate plants) is also referred to as hashish, grass or marijuana, and is generally inhaled. It can lead to a dependence syndrome, as well as mild cognitive impairment. The main active component is tetra-hydro-cannabinol, which affects the cannabin receptors.

(2) Marijuana cigarettes are sometimes treated or saturated with PCP, which may cause PCP over dosage. Hashish oil can be used the same way. Since the drug has a high potential for abuse, it is closely regulated under the Federal Control Substances Act.

6. LSD or lysergic acid diethylamide (a crystalline acid from ergotic alkaloids) produces in the mind vivid hallucinations. (2) Significant unfavorable reactions by LSD include prolonged, delayed, and recurrent reactions such

as depression and long-term schizophrenic or psychotic reactions. A bad trip on the drug is likely to be a paranoid experience, and tendencies towards violence can be characteristic of LSD intoxication.

7. Phencyclidine (PCP) is a hallucinogen. It was developed as an anesthetic for dissociate anesthesia, which puts the person under its influence in a cataleptic state, where he appears to be awake but is detached from the surroundings and unresponsive to pain. (2) PCP has a history of leading to the most serious adverse effects resulting in more suicides, assaults, and murders from its' usage than any other addictive drug. Prolonged periods of psychosis in even normal persons have been reported.

Source: (1)News and Views Feature – Drug Addiction: Bad Habits Add Up By Trevor W. Robbins and Barry J. Everitt, who are in the Department of Experimental Psychology, University of Cambridge, Downing Street, Cambridge, UK – Copyright Macmillan Magazines Ltd., 1999. (2)Mosby's Pharmacology in Nursing, 18th edition, by Leda M. McHenry, Ph.D. R. N., Faculty University of Massachusetts, Amherst, MA., and Evelyn Salerno, Pharm.D., R.Ph. Adjunct Professor, University of Miami School of Nursing, Miami, Florida – Publisher Mosby Year Book, 1992, Pages 169 to 174

Note: I know the above is somewhat technical and difficult to understand unless you are an RN or MD or chemist. I did my best to make it readable for the average person, and hopefully you will see around the technical stuff, and get the meaning and more important, the message. The message is one of life and death: these drugs are detrimental to the brain's operation and of course extremely addictive. They aren't recreational and once on a track of using them can change a person's life and happiness. Stay away from them, no matter how cool your peers say they are! Remember there is nothing cool about addiction. Go to a rehab center and observe these poor people, who are under its' power, if you don't believe me.

Is there a cure for addiction? Yes and no. There is no cure for the afflicted person, unless he wants to get better, and is willing to admit he has a

problem.   Other than that, medical science just doesn't have an answer to the disease.   Yes there are medicines and rehab centers, which help, but they don't cure!   But
all isn't lost for the addicted.   There is hope, if he wants to reach out for it.   I'll discuss this in the next chapter.

(Note: For another opinion concerning a cure for addiction, refer back to Bad Diet Linked to Alcoholism, Page 57.)

CHAPTER FORTY
ALCOHOLICS ANONYMOUS

Alcoholics Anonymous (A. A.) was co-founded by William Griffith Wilson (Bill W.) and Dr. Robert Holbrook Smith (Dr. Bob) on June 10, 1935. Wilson conceived the idea of the organization, while hospitalized for excessive drinking in December of 1934. During his stay there, Wilson received a spiritual experience that permanently cured his desire to drink. Armed with this experience, he began a crusade to persuade other alcoholics to stop drinking, as he had. Wilson's first convert was Dr. Smith, who agreed to adopt his method to find freedom from alcoholism. Four years later, the two published the book, Alcoholics Anonymous, which contains the Twelve Steps and a spiritually based program of recovery for alcoholism.

A. A. was established "to help the sick alcoholic recover, if he wishes". Regular meetings are held by about 12,000 Alcoholic Anonymous groups located in both the United States and abroad. World membership in 2003 came to 2,160,013. The only requirement for membership is for the alcoholic to admit that he is one, and to express a sincere desire to stop drinking. Members are required to follow the A. A. program's twelve steps of recovery, which are based on principles of human conduct found in medicine, psychiatry, religion, and in their own experience. The organization is supported by voluntary contributions from its members with a board of trustees (the Alcoholic Foundation), which serves without compensation. Headquarters are in New York City.

THE TWELVE STEPS OF ALCOHOLICS ANONYMOUS

1. We admit we are powerless over alcohol – that our lives had become unmanageable.
2. Came to believe that a Power greater than ourselves could restore us to sanity.
3. Made a decision to turn our will and our lives over to the care of God as, we understand Him.
4. Made a searching and fearless moral inventory of ourselves.
5. Admitted to God, to ourselves, and to another human being the exact nature of our wrongs.
6. Were entirely ready to have God remove all

these defects of character.

7. Humbly asked Him to remove our shortcomings.

8. Made a list of all persons we had harmed, and became willing to make amends to them all.

9. Made direct amends to such people wherever possible, except when to do so would injure them or others.

10. Continued to take personal inventory and when we were wrong, promptly admitted it.

11. Sought through prayer and meditation to improve our conscious contact with God, as we understand Him, praying only for knowledge of His will for us and the power to carry that out.

12. Having had a spiritual awakening as the result of these steps, we tried to carry this message to alcoholics, and to practice these principles in all our affairs.

"The steps of a good man are ordered by the Lord: and he delighted in his way."
Psalm 37:23

Source: The Twelve Steps for Christians Revised Edition Based of Biblical Teachings Friends in Recovery RPI Publishing, Inc., 1725 Kresky Avenue, Centralia, WA 1994, Pages xi– xix

CHAPTER FORTY ONE
MANIFEST DESTINY

If the term, "All roads lead to Rome," applied to the Roman Empire; then the term, "All inventions of major importance come from the United States," should apply to the American Empire. Think about it: everything I wrote about concerning environmental problems came out of inventions from the United States i.e. autos, nuclear energy, mass farming techniques, abortion, forest depletion, the Colt revolver hand gun, organized crime, nuclear weapons of mass destruction, etc. Going on with the facts of the data collected in the book, I went a step further and asked the question, why. Why was this done? What was the motivating force, which brought it to a worldwide dominating stature?

I believe it was accomplished by 'Manifest Destiny'. (1)Manifest Destiny is a term implying divine sanction for the United States territorial expansion. It was first coined in the 1845 issue of the United States Magazine and Democratic Review, edited by John L. O'Sullivan.

"Nothing must interfere," he wrote concerning the growth rate of the new USA, "with the fulfillment of our 'manifest destiny' to overspread the continent allotted by Providence for the free development of our yearly multiplying millions."

President James K. Polk, the country's 11th President (1845 – 1849) would use this theme, making himself the spokesman of American expansionism. Under his administration, the country would add Texas and Oregon territories, and the great southwest. These lands comprised the entire western part of the United States. He did all this, as he campaigned to do, in only one term as President.

My feeling is 'Manifest Destiny' didn't end with President Polk's expansionism of 1846. I believe it has continued up until the present time, influencing every corner of the earth; doing so with a (2) mechanized military force unequalled in the world:

1. The U. S. has the largest Air Force: 4,413 combat aircraft as of September, 1999. This total includes 179 bombers, 1,666 fighter and attack aircraft, and 1,279 trainer aircraft.

2. The U. S. Navy, at the start of 2002, had the most aircraft, more than 4,000 jet planes in service. Included in this are about 20 different models of fixed-wing aircraft, with carrier-based F-14 Tomcat and F/A-18 Hornet fighters and various models of reconnaissance, transport, anti-submarine, and airborne command post aircraft. The Navy also operates six different helicopters and the unique new V-22 Osprey tilt rotor aircraft.

3. The U. S. Navy has in service 74 armed submarines, more than any other country. Their submarine fleet is nuclear powered and possesses more than 50 Sea wolf and Los Angeles class attack ships, as well as 18 Ohio class ballistic missile carriers.

4. In 2002 the U. S. Navy had the largest number of warships: 318 principal vessels, including submarines. These ships are supported by 380,000 Navy personnel and 183,000 civilians.

5. The warships with the largest full load displacement in the world are the Nimitz class US Navy aircraft carriers USS Nimitz, USS Carl Vinson, USS Dwight D. Eisenhower, USS Theodore Roosevelt. USS Abraham Lincoln, USS John C. Stennis, USS George Washington,

USS Harry S Truman and the USS Ronald Reagan; the last five of which displace about 217.2 million lbs. (98,550 tonnes). The ships are 1,092 ft. (332.9M) long, have 4.49 acres (1.82 HA) of flight deck, and driven by four nuclear-powered 260,000-hp (194,000-kW0 geared steam turbines, and can reach speeds of over 30 knots (34.5 mph or 56 km/h).

Note: All of the above and a well trained standing Army are supported by the world's largest military budget of $396.1 billions dollars.

Source: (1)The American Nation, A History of the United States, by John A. Garraty, Columbia University, A Harper-American Heritage Textbook, Harper & Row, Publisher, Inc., New York, NY, 1966, Page 312

(2) points 1 through 5, Guinness World Records 2002, World Copyright Reserved, Copyright 2003 Guinness World Records Ltd., Page 152

I sometimes allow myself to imagine: what if our leadership here in America could put the 'Manifest Destiny' drive on hold for about five or ten years and use some of the almighty defense budget to rebuild our country's roads and bridges, of which the nation is in desperate need of replacing. Perhaps the American tax payer would for once come out ahead. When this is done, America can go back to controlling the world again.

America's war veterans haven't died in vain, protecting 'Manifest Destiny'. Every war fought by the United States has left the American economic stamp on it; not only changing that country's economy, but their culture as well. Starting with President Polk's Mexican War:

Mexican War (1846 - 1848) - It is obvious, America coming out of this war, stole the best land from Mexico. But it did more than that, it left an economic presence, which relegated Mexico to being a third world country, cutting them off from the commerce of the mainland United States.

Civil War (1861- 1865) - Before the Civil War the United States wasn't actually united. They were two countries: the North and the

South, with two different cultures. The War Between the States settled this: the South would be like the North and comply with 'Manifest Destiny'.

Spanish-American War (1898 - 1902) - This placed the U. S. in control of Latin countries in America; not actually taking the countries over, but showing them clearly, who's boss in the Americas. If US companies decided to do any business with the Latinos, no one there should interfere with their progress.

World War I (1917 - 1918) - Actually World War I never ended: it simply turned into World War II - the European side of the war. And this is how I would like to treat it, as one war. The result was, when all the killing was over, the U. S. would step in with military occupation, and leave its' economic recovery program in place. This meant big bucks for America's big business.

The second side of World War II (1940 - 1945) was in the South Pacific against the Japanese; like the European side, the U. S. would militarily occupy Japan and impose not only their economic system there, but change their culture as well. This was the first American 'Manifest

Destiny' foothold on the Asian continent.

Korean War (1950 – 1953) – This war broadened the American grip onto the Asian continent and in South Korea would establish the economic model that would stand as a convincing inroad of U. S. commercial might.

Vietnam War (1964 – 1975) – Very similar to the Korean War, but would make an impact in Southeast Asia and establish 'Manifest Destiny' there going further west into India and Pakistan and as far south as Australia.

Gulf War (1990 – 1991) – Relatively a minor war as American wars go, but it would make the U. S. presence in the Middle East known and our desire to defend our oil interests in the area.

Iraq War (2002 – present) – Or President George W. Bush's War, as it would later be called, was simply a second phase of the Gulf War with the same goals.

EPILOGUE

My short environmental journey has ended, leaving me in shock. I never realized how bad the environment has become in the short span of my lifetime. The research I did, came from readily available published material, such as the Guinness World Records, old history text books, the internet, Time Almanac, television, encyclopedia and others. So all this information wasn't hidden; anybody, who was interested, could easily find it, if they were inclined to do so. I simply tapped into the various sources, and put it all down compactly in this little book, trying to make it understandable for the regular person. Shocking; it is indeed shocking. It is my hope that you the reader will likewise be shocked by its contents and do something about it. It is you the people, who can turn this around. Those in power know this, and know if they can keep you in ignorance, they will continue to take the easy road of doing nothing, and stay rich and in power. They could care less about the planet we live on and our quality of life. The book my partner and I wrote, if you take it to heart, will expose the ignorant power brokers, who are slowly destroying our planet. The work insists that these polluters explain themselves for their deception and demands that they look for ways of cleaning up their product. We the people aren't at their mercy, if we act. Pressure can be administered through citizen's rights groups and a caring electorate, who are actively petitioning their congressmen. The government is equally guilty concerning the problem, which permits a large monopoly oil companies to not only exist but to go unchecked. Trusting oil executives to clean up their product on their own is almost a laughing matter. However, none of this is humorous. Don't let it come down to this. No you can't trust monopolies to do the right thing. They don't care about ethics; they only care about making profits. And they will make them no matter what it takes to do it. Don't let them get away with this! The environment is your business. Get involved. It isn't too late, but it will be, if something isn't done. It is time!

Saying all that, I would have ended the book right here, if it wasn't for one thing, which stood out throughout the writing: all pollutions of major concern originated from inventions coming out of the United States. An example is the assembly line manufacture of the automobile by its' inventor Henry Ford.   By far the most threatening of all the pollutants to life on earth, comes out of the combustion engine car, as was earlier documented. There are other examples, of polluting inventions, as you will recall from the details of the book. Through 'Manifest Destiny' these inventions were taken all over the world. It isn't our purpose to malign these inventions, since they are worthwhile and essential to man's lifestyle. But they are polluters and as such should be cleaned up as much as the state of technology will allow. Getting back to the automobile, it is indeed our 'bread and butter'; without transportation the wheels of commerce simply do not work. Although this is true, there are alternatives to the polluting fossil fuels of today's cars, such as natural gas and other fuels. These natural gas cars are just as efficient, and can be assembly line manufactured if the fuel for them was made available at local gas stations. It isn't and won't be. Why?   Ask the monopoly controlled oil industry leaders. Simply, they don't like the competition!   That's the end of that.

It isn't the end of the problems created by the stubbornness of this close minded mentality of the oil industry. Up until this point, our book said nothing about the dependency on foreign oil, but we think, we should even though it isn't directly related to the environment. Most of America's oil needs are met by oil producing nations, which directly affects our balance of payment. Without going into detail, these purchases are quite large, sending money abroad, which could better serve our nation, if used domestically i.e. more jobs, a more stable dollar in relation to foreign currency, etc. It is a debt incurred by our nation that has to be paid, and we are paying for it every day with higher prices at the pump. Prices are dictated by a large international oil cartel. Along with this problem, it is compounded by a growing involvement our nation has with these oil producing countries' internal political affairs, which threatens their oil production. Hence, the two Persian Gulf wars. The oil companies and the government's answer to this is: to dig more wells domestically. This answer is short term and insipid. It doesn't solve the environment problem; it makes it worse. My partner and I have given practical energy alternatives, which would greatly reduce our nation's dependency abroad. If we knew about them, why didn't they?

To add to the environmental dilemma America has a government, which is slow to react to the situation, and an electorate equally slow to get involved. And although I'm making an appeal to the people, I have little hope that it will do any good. I say this with an apology to those of you who do care, but regretfully you aren't the majority. The majority continues on the same polluting path, poisoning themselves and their children and killing the wonderful planet we all live on and depend upon. All of it is insane; why do they do it? The book would lack credence, if I didn't at least try to answer this question. I tried by studying Americans. The work was an adventure and got me into many arguments, because Americans don't like being studied. They become defensive, and it is a chore to gain rational information. You can't write if you don't have anything to write about, so I went ahead anyway. This is what I found out.

There is something deeply entrenched in the mindset of the American people and their

leadership. The pollution goes on because the people allow it. There seems to be an evil inside the entire American system, which I can't explain. The typical American has little morality in him. He feels entitled to anything he wants, whether he has a right to it or not. He hates deeply, especially other Americans. He is driven by this hatred, even to the point of killing another American for a minor offense, like beating him out on the highway. With this hatred, there is a know-it –all mentality which insists he knows everything about everything. This know it all attitude closes his mind to learning. What he has learned up until a certain point in his life, is all he'll ever know. This ignorance and hatred seems to fit into the American system and limits it from doing anything worthwhile. The facts gathered in our book concerning the environment appear to support this thesis. To say that America is an evil empire maybe premature at this point. However, the facts supporting this book suggest otherwise.
This harsh attitude prevalent in the nation isn't limited to its' borders, but has grown and spread all over the world. There isn't a corner of this globe, which isn't manipulated by the bullying

tactics of the USA. If she doesn't beat a country militarily, she can take it out on them economically. In short, America rules the world. It is as if they have a gun pointed at the world. Anybody who steps out of line will be shot.

So how did this mind set come to be? Did the empire have a curse placed on it to be unethical, or did it just evolve that way? Do such major events of this magnitude just happen? We don't believe they do. A power greater than mankind directs the happening. Can it be proven? We believe it can, but that is a thesis for another book. My partner and I feel this proof of ours, which we call The Fifth Vision, wouldn't be appropriate for this writing, since it is based on Biblical prophecy. Consequently, it will appear in another book of ours, entitled: The Spectral. This second book is the antithesis of Everything You Should Know, etc. in that its' theme is based on the spiritual and the unexplained. What we call: "the other environment."

END

EVERYTHING YOU SHOULD KNOW ABOUT THE WORLD'S ENVIRONMENT, BUT ARE INDIFFERENT TO ASK!

BIBLIOGRAPHY (references)

Good News Bible – Catholic Study Edition – Sadlier, A division of William   H. Sadlier, Inc., New York, Chicago, Los Angeles –

Guinness World Records 2002 & 2003 – Gullane Entertainment – A Gullane Entertainment Company World, Copyright 2003 Guinness World Records, Ltd.

TIME Almanac 2003 with Information Please – Information Please, Boston, MA – part of Family Education Network, Inc. – TIME Inc. Home Entertainment – Copyright 2002 by Family Education Network, Inc.

The Encyclopedia of Common Diseases – by the Staff of Prevention   Magazine – Rodale Press Inc. – Copyrighted 1976 by Rodale Press, Inc.

The Twelve Steps for Christians, Revised Edition – Based in Biblical Teachings – Friends in Recovery – Published by RPI Publishing, Inc., Centralia, WA – Copyrighted by Friends in Recovery 1988, 1994

Mosby's – Pharmacology in Nursing – by Leda M. McKenry, Ph. D, R. N. and Evelyn Salerno, Pharm., R. Ph. – Mosby Year Book, Dedicated to Publishing Excellence, St. Louis, Baltimore, Chicago,   London, Philadelphia, Sydney, Toronto – Copyrighted 1992 by Mosby – Year Book, Inc.

Stedman's Concise Medical Dictionary Illustrated, Second Edition – Williams & Wilkins, Baltimore, Philadelphia, Hong Kong, London, Munich, Sydney, Tokyo – a Waverly Company – Copyrighted 1994

The World Almanac and Books of Facts, 1999, Copyrighted 1998 by PRIMEDIA Reference, Inc. – World Almanac Books,

Mahwah, New Jersey – a PRIMEDIA Company

The American Nation – a History of the United States by John A Garraty –
Columbia University – Harper & Row, Publisher, Incorporated, American Heritage Publishing Co., Inc., New York and London – Copyrighted 1966

Funk & Wagnall's New Encyclopedia – Funk & Wagnall's, Inc., New York, NY – Copyrighted 1971, 1972, 1973

The Glucose Revolution Life Plan by Jennie Brand-Miller, Ph. D, Johanna Burani, M.S, R.D.,C.D.E. & Kaye Foster-Powell, B.SC., M. Nutrition & Diet. – Published by: Marlowe & Company, an Imprint of Avalon Publishing Group Inc., New York, NY – Copyright 2000, 2001

LIFE Sixty Years – a 60th Anniversary celebration 1936 – 1996 – By The Editors of Life, Published by LIFE Books, Time Inc., New York, NY Copyrighted 1996, Time Inc. Home

Entertainment

Economics – An Introductory Analysis, by Paul A. Samuelson, Professor of Economics, Massachusetts Institute of Technology, Publisher, McGraw-Hill Book Company, Inc., New York, NY, 1961

Isaac Asimov's Book of Facts, Bell Publishing Company, New York, NY,

Copyrighted by Red Dembner Enterprises Corp. 1981 Edition, Distributed by Crown Publishers, Inc.

Environmental Law Examples and Explanations, Steven Ferry, Aspen Publishers, 3rd ed. 2004)

Webster's Seventh New Collegiate Dictionary, A. Merriam-Webster, G. & C. Merriam Company, Publishers, Springfield, MA 1963

Information Provided Off the Internet:

...Food Additives – Protect Your Family From Cancer Causing Chemicals! Copyright 2003-2007, Healthy Eating Advisor.com – All right reserved. The Healthy Eating Advisor, Escondido, CA

...Radioactive Wastes: WNA 11/1/08 Copyright World Nuclear Association,
All right reserved. 'Promoting the peaceful worldwide use of nuclear power as a sustainable source.'

...No Quick Fix for the Ozone Hole / Live Science, By Life Science Staff,
Posted: 30 June, 2006, Page 1.

...NASA Ozone Hole Watch: What is the Ozone Hole? October 4, 2004, NASA Official: Paul Newman

...Food Additives to Avoid – Dr. Janet Starr Hull's newsletter, Creator of the Aspartame Detox Program, Aspartame Dangers Revealed /

Disclaimer / Link to us / Contact / Site Map / Search Copyright 2002, Sweet Poison.com. All right reserved.

...Metabolic Syndrome: Risk Factors – MayoClinc.com – Copyright 1998-2008 Mayo Foundation for Medical Education and Research (MFMER)

...Health Care in the United States – Wikipedia, the free encyclopedia
Wikipedia is a registered trademark of the Wikimedia Foundation, Inc., a U. S. registered tax-deductible nonprofit charity.

...Health Care Costs – a Primer – August, 2007 – Key Information on Health Care Costs and Their Impact   T Henry / J. Kaiser Family Foundation.

...Why does U. S. Health Care Cost So Much? (Part I) – Economic Blog –   NY Times. com – Excerpts from the book, Why Does U. S. Health Care Cost So Much?
by Uwe E. Reinhardt, Economist at Princeton

University – December 24, 2008.

...OECD Health Data 2007 from the OECD Internet Subscription Database updated July, 2007. Copyright OECD 2007.

...News and Views Feature – Drug Addiction: Bad Habits Add Up – by Trevor W. Robbins and Barry J. Everett, who are in the Department of Experimental Psychology, University of Cambridge, Downing Street, Cambridge, UK – Copyright 1999 MacMillan Magazines Ltd.

...Depression: Signs, Symptoms, Types and Ways to Get Help – Joanna Saisan, MSW, Melinda Smith, M.A., Robert Segal, M.A, and Jeanne Segal, Ph. D, contributed to this article. Last modified in July, 2008

...What is Depression? – PRISTIQ.com – Copyrighted 2008, Wyeth Pharmaceuticals Inc.

...Health – Vital Information with a Human

Touch, Topic Overview –

Health.com – Bipolar Disorder – Copyrighted 2008 – Health Media Ventures, Inc.

...Anxiety Symptoms, Causes, Types, Signs and Treatment on
MedicineNet.com – Copyrighted 1996–2008
Medicine Net, Inc. All rights reserved.

...Supreme Court Judgment on Noise Pollution, November 28, 2007, CIMOGG (Citizens' Movement for Good Governance) Articles and Publication, Page 1 & 2

...World Disasters. Weather Related Timeline, 21 Century, World
Weather Related Timeline, March 29, 2009, Page 1 to 5, Copyrighted CNT Groups, 2000.

...Radioactive waste – Wikipedia, the free encyclopedia, Page 1–12, November 1, 2008.

...Cyclone Nargis – Wikipedia, the free encyclopedia, Wikipedia is a registered trademark of the Wikimedia Foundation, Inc., a

U.S. registered 501(c) Tax-deductible nonprofit charity. March 29, 2009

...Natural Hazards - Pestilence & Disease, Page 1, NDA Natural Disaster Association, Information Resource on Natural Hazards & Disasters, March 28, 2009

...U. S. Environment Protection Agency, Office of Mobile Sources, EPA 400- F-92-007, Fact sheet OMS-5 August, 1994, Automobile Emission: An Overview, Pages 1 to 4

...Also Automobiles and Ozone EPA 400-F-92-006, Fact Sheet OMS-4, January, 1993, Pages 1 to 6.

TABLE OF CONTENTS

Preface   pages 3-5

Chapter 1   Introduction   p. 6-9

Chapter 2   Concepts of Ecosystems p.10-12

Chapter 3   Disasters (worldwide) p. 13-17

Chapter 4   Oil Spills p.18-20

Chapter 5   Nuclear Power Plants' Accidents p. 21-23

Chapter 6   Henry Ford (1863 - 1947) Founder of Ford Motors - And What Has Been Done to Clean Up His Car?   p. 24-30

       ...Evaporative Emissions

       ...Refueling Losses

       ...Exhaust Emissions

       ...Clean Air Acts

Chapter 7   Major Pollutants   p. 31-34
    ...Ozone
    ...Carbon Monoxide
    ...Nitrogen Dioxide
    ... Greenhouse Gases – Carbon Dioxide, Methane & Nitrous Oxide

Chapter 8   Global Warming   p. 35-42
    ...Climate Changes – Storms, Floods
    ...Dry Climate Changes – Droughts
    ...Pestilence – AIDS, Malaria, Tuberculosis

Chapter 9   Radioactive Strontium 90 & Bone Cancer   p. 43-46

Chapter 10   Cancer (definition)   p. 47-51
    Lung Cancer from the Air We Breath
    ...Exercise & Cancer

Chapter 11   Ozone Hole   p. 52-54

Chapter 12   Mars' Disaster (Could it happen on Earth?)   p. 55-61
    ...Similarities to Earth

...Pathfinder Lander & Sojourner Rover

...Water Discovery on Mars

Chapter 13   Photosynthesis   p. 62-66

...Amazon River

...Rain Forest Deforestation

...Acid Rain

Chapter 14   Nuclear Energy   p. 67-69

... (how it came about as an Energy source)

Chapter 15   Nuclear Waste (Radioactive) p. 70- 74

...Department of Energy

...Nuclear Waste Treatment & Management

...Disposal of High Level Wastes

Chapter 16   Vested Interest   p.75-81

Chapter 17   Solutions (10 winners) p.82-86

Chapter 18   The Swelling World's Population
           p. 87-92
           ...Poverty
           ...Overpopulation
           ...Law of Diminishing Returns

Chapter 19   Abortion – Definition   p. 93-96
                    ...Statistics

Chapter 20   Abortion Reality – Genocide?
           p. 97-99
        ...Death as a Means of Controlling
           Population
        ...Photos of Eight Weeks Gestation

Chapter 21   Genocide – American Indian
           p. 100-108
           ...History   ...
           ...What's become of the
              American Indian?

Chapter 22   American's Eleven Most
           Endangered Rivers p. 109

Chapter 23   Fresh Water – Introduction
             p. 110–116
             ...Provide for More People
             ...Irrigation
             ...Amount of Water
             ...Israel – Water Pipelines

Chapter 24   Sahara Desert & The Nile River
             p. 117–122
             ...Description of the Sahara Desert
             ...Path of The Nile River

Chapter 25   Water Pollution by People p.123

Chapter 26   Food & Diet   p. 124

Chapter 27   Bad Diet Linked to Alcoholism
             p.125–130
             ...Various Supporting Research

Chapter 28   Food & Drug Administration
             p. 131–134
             ...How it was organized ?
             ...Its' responsibility
             ...Adulteration

Chapter 29   Food Additives   p.135-148
       ...Cancer Causers
       ...List of Common Additives

Chapter 30   Metabolic Syndrome- Definition
       p. 149-153
       ...Factors - Age, Race, Obesity, History
       ...Other Diseases

Chapter 31   The Glycemic Index   p.154-158
       ...High GI Foods (bad)
       ...Low GI Foods (good)

Chapter 32   Other Food Information p. 159-162
       ...Smoked Meats/Fishes

Chapter 33   Nitrites p. 163-167
       ...How They Work in the Stomach.
       ...Cancer Causer
       ...Makes Meats Look more Appetizing

.

Chapter 34   Mortality in the United States
       p. 168-169
       ...Number/Type of Death

Chapter 35   Depression – Definition

        p. 170-177

        ...Everyday Depression (beginning)

Chapter 36   More Depression p. 178-186

        ...Everyday Depression (continued)

    By Dealing with People, Bad Laws,

Depressing History, No Road Courtesy, etc.

Chapter 37   Medical/Health Care Insurance

        Complex p. 187-197

    (part of More Depression)

    ...Statistics: comparison to other

        Developed Countries

...Price of Prescription Drugs

...Delays in Seeking Medical Care

...Why American Health Care is So High.

Chapter 38   Depression For No Reason:

        p. 188-207

    ...Bi Polar Disorder

    ...Generalized Anxiety Disorder

...Suicidal Depression

...Post partum Disorder (after birth)

...Alcohol & Drugs

Helpful solutions:

...Develop support Systems

...Life Style Changes

...Medication

Chapter 39   Addiction   p. 208-217

...Type of Person Addicted

...Common Drugs of Addiction

Chapter 40   Alcoholics' Anonymous p. 218-220

... Bill Wilson's Twelve Steps

Chapter 41   Manifest Destiny   p.221-227

...America's Territorial Expansion

...American Armed Forces (statistics)

...Wars Involving the United States

Epilogue   p. 228-235

   ...Reason for Writing the Book
   ...Plea for Americans to get Involved
   ...American Mindset
   ...Conclusions About the Environment
   ...American Empire Rules.

www.ingramcontent.com/pod-product-compliance
Lightning Source LLC
Chambersburg PA
CBHW051801170526
45167CB00005B/1831